Astronomers' Universe

Series Editor
Martin Beech
Campion College
The University of Regina
Regina, Saskatchewan, Canada

The Astronomers' Universe series attracts scientifically curious readers with a passion for astronomy and its related fields. In this series, you will venture beyond the basics to gain a deeper understanding of the cosmos—all from the comfort of your chair.

Our books cover any and all topics related to the scientific study of the Universe and our place in it, exploring discoveries and theories in areas ranging from cosmology and astrophysics to planetary science and astrobiology.

This series bridges the gap between very basic popular science books and higher-level textbooks, providing rigorous, yet digestible forays for the intrepid lay reader. It goes beyond a beginner's level, introducing you to more complex concepts that will expand your knowledge of the cosmos. The books are written in a didactic and descriptive style, including basic mathematics where necessary.

More information about this series at http://www.springer.com/series/6960

Pierre Léna

Astronomy's Quest for Sharp Images

From Blurred Pictures to the Very Large Telescope

 Springer

Pierre Léna
Observatoire de Paris & Université de Paris
Meudon, France

ISSN 1614-659X   ISSN 2197-6651   (electronic)
Astronomers' Universe
ISBN 978-3-030-55810-9   ISBN 978-3-030-55811-6   (eBook)
https://doi.org/10.1007/978-3-030-55811-6

Cover illustraton: The full Moon sets at dawn, behind the domes of the Very Large Telescope distributed over the Cerro Paranal platform in the Atacama desert (Chile). The photography is taken 14 km away from Paranal, on the road to the future European-Extremely Large Telescope at Cerro Armazones. Credit: G.Gillet/ESO

This Springer imprint is published by the registered company Springer Nature Switzerland AG.
The registered company address is: Gewerbestrasse 11, 6330 Cham, Switzerland

To Sang Shao-Hua
To the memory of Lodewijk and Ulla Woltjer

Preface to the French Edition

We are such stuff as dreams are made on—William Shakespeare, *The Tempest*

I have always been very lucky! Starting out as a professional astronomer at the beginning of the 1960s, my life in research and teaching, which has now reached its end, encountered so many extraordinary developments in astrophysics over the past half-century. There were so many different telescopes to visit and use! So many hi-tech instruments, designed by generations of enthusiastic students and researchers! And so many improvements brought about by the rise of information technologies! I have experienced all that. I have spent many long nights gazing at the sky. I have witnessed magnificent discoveries, thanks to these observatories on Earth or in space. From Europe to America, from Chile to China, I have been able to gauge the universal nature of science and it has been my pleasure to belong to the community of all those curious beings who serve it with such enthusiasm. I have also experienced the potential of a unified Europe and the rich rewards that can spring from it.

The field my own work may have contributed to, and which the work of the students and teams around me certainly did, has become truly immense. It is becoming more productive by the day and now employs hundreds of researchers and engineers. And so it seemed to me important to tell this story of women and men, mirrors, fringes, and stars and to share my inside view as a privileged witness and modest actor.

I am going to tell you a story about blur, a picture book of sorts, but told with words. It is a quest that began with Galileo, one night in Venice to be exact, in 1609, when he pointed his refracting telescope toward the Milky Way, then Jupiter, Saturn, and Venus, and even the Sun itself, discovering everywhere images that no one had ever seen before him. Blurred and imprecise images which immediately brought about a scientific revolution. From that point, astronomers would make constant progress, discerning ever more detail in sharper and sharper images, first peering wide-eyed into refracting and reflecting telescopes, then studying photographic plates, which can show so much more, and finally sending unmanned probes to Mars or Comet 67P/Churyumov–Gerasimenko to photograph these distant bodies at point-blank range.

Between Galileo and the 1960s, which were precisely the years when I began my life as a research scientist, progress was certainly made in the quest for finer detail, some of it very significant. However, this progress did not concern the wavelengths at which stars and planets emit most of their light, namely visible and infrared wavelengths. For these lights, there has been almost no progress! In Arizona, as a young researcher, I was already confronted with the blur in images of the Sun. After some adventures that were as much aeronautic as astronomical, I only returned to the problem of imaging a decade later, inspired by Antoine Labeyrie and also by François Roddier, two exceptional physicists of my own generation. I then had the good fortune to be associated with the design of the Very Large Telescope, now set up in Chile, between 1978 and 2005. Thanks to the trust shown to me by Lodewijk Woltjer, director of the European Southern Observatory (ESO), of which France was a member, I found myself at the heart of a new campaign against the blur in astronomical images.

With hindsight, I am amazed to see how productive those years were. I always felt as though I was making my way through a thick fog, never sure whether I might stumble at the next step. So where did I find the strength to carry out this struggle with myself, with the reluctant instrument, with the sceptic, and with those who would ironise? This book tells that story. I owe much to my family, to friendships and emotions, to a long scientific tradition which I had inherited without even realising it. This story is my way of describing and giving thanks to these legacies.

I am aware of having been just a relay, just another link in the chain. I often mention the young students who joined me on this journey, and I have tried to show how there is a continuity from generation to generation, something I am very conscious of. These are the ones who have made the extraordinary discoveries I describe in this book, who are engaged in still further attempts to reduce the blur, and who are preparing the future.

And of course, this book is also dedicated to them.

Acknowledgements

My thanks go first to Antoine Mérand and Andrés Pino for their invitation to give a talk on the history of the VLT, through my own personal experiences, during a trip to Paranal in 2016. That was where the idea of this book was first formulated, and my wife Sang Shao-Hua was behind it all the way. She deserves the dedicacy. Thérèse and Xavier Perras provided me with precious moments in their 'writing retreat' in Burgundy.

Progress would have been difficult without discussions with the many people who played their own role in this story, including in particular Daniel Bonneau, Vincent Coudé du Foresto, Pierre Cox who invited me to ALMA, Françoise Delplancke, Frank Eisenhauer, Reinhard Genzel, Andreas Glindemann, Daniel Hofstadt for his warm welcome at Rupanco Lake, Pierre Kervella, Bertrand Koehler, Antoine Labeyrie, Anne Lagrange for our meetings in Chartreuse, Denis Mourard at Calern, Norbert Hubin, Thibaut Paumard, and Guy Perrin for reading the manuscript, Yves Quéré, Daniel Rouan, Jean Schneider, Farokh Vakili, and Julien Woillez: I am sincerely grateful to all of them. Thanks also to Pierre Chavel, for supplying me with genuine fringes. Many others are cited, but there just was not time to exchange recollections with everyone. Many thanks also to them.

Using my professional archives at the Paris Observatory, I was able to recover all the necessary information, thanks to the excellent organisation of Nicole Fouquet, who was an invaluable secretary throughout these decades. For the classification of the archives, I owe everything to Marie-Agnès Dubos, while Agnès Fave helped me to access them. I am truly grateful to all three.

Without the encouragement of Sophie Bancquart and the careful editing work of Juliette Thomas, I doubt whether this project would ever have been completed. A huge thank you to my French publisher *Le Pommier*!

Note on the English Edition

I am very happy that Springer accepted to publish this translation of the book which came out under the title *Une Histoire de Flou* in France at the beginning of 2019. I thank Luc Dettwiller for his careful reading and corrections, and the colleagues who provided illustrations. I would particularly like to thank the editor Ramon Khanna for his support and the translator Stephen Lyle for the quality of his work, as well as the editing team in India. My warmest thanks go also to Xavier Barcons, Director General of the European Southern Observatory, for the help with this edition which he agreed immediately.

Since I wrote the book in 2018, adaptive optics and interferometry have been bringing in ever more outstanding results. I have only been able to mention a few of them rather briefly in the present publication, but they confirm the victorious trend up until now in the war on blur, if such was necessary.

I am particularly happy to share this story with all those who may not read French but have taken part in this great adventure over the past half-century or are contributing to it in the present century. I hope I can be forgiven for my deliberately personal account of this, with all its omissions and sometimes personal assessments, for which of course I assume responsibility.

On October 8, at the very moment when this book goes into press, the Nobel prize in physics is given to Reinhard Genzel and Andrea Chez, for their work on the black hole at the galactic center, and jointly to Roger Penrose for his theoretical work on general relativity and black holes. As this is an extraordinary step in the story I have attempted to tell, it should not be omitted here.

Lo and Ulla Woltjer, to whom I owe so much, have both recently departed this world. I dedicate this English edition of the book to them.

Paris, France   Pierre Léna
October 2020

Contents

1

One Night in Paranal

And from that primal night in which two men born blind grope for their ways, the one equipped with the tools of science, the other helped only by the flashes of his imagination, which one returns sooner and more heavily laden with a brief phosphorescence? The answer does not matter. The mystery is common to both.—Saint-John Perse, *Speech upon receiving the Nobel Prize in Literature*[1]

In these latitudes, night falls quickly. At midday, the Atacama desert in Chile, drenched in heat under the almost vertical noonday sun, loses its shadows, its relief, and its colours, only to get them back at the close of day. To the west, a veil of white hugging the horizon has now become a layer of nearby cloud, and beneath this, a few kilometers away but invisible, lies the Pacific Ocean. Far away to the north-east, the peak of the volcano Licancabur (5916 m) and the snow-topped Andes form the horizon, above which the dark shadow of the Earth is now climbing.[2] At an altitude of more than 2600 m, I am alone up here, and I feel totally alone, surrounded by the silence of the desert, as I await the onset of night. Above me, an immense sky turns dark blue and deepens, and the wind has dropped. On the vast horizontal platform that surrounds me stand a group of strange buildings. Four of these, almost standing in a row, are truly gigantic metal structures, catching the last reddening rays of the

---

[1]www.nobelprize.org/prizes/literature/1960/perse/speech/.

[2]This anti-twilight arch is also known as the Belt of Venus. See Lynch and Livingston (2001).

© Springer Nature Switzerland AG 2020
P. Léna, *Astronomy's Quest for Sharp Images*, Astronomers' Universe,
https://doi.org/10.1007/978-3-030-55811-6_1

Fig. 1.1 On the VLT platform, looking to the twilight at West, with three egg-cup shaped telescopes and three astronomers preparing for the night. A conjunction of three planets can be made out in the twilight sky, just to the right of the rightmost dome: Jupiter at the top, Venus bottom left, and Mercury bottom right. Credit: ESO/G. Brammer

setting sun. Four others, smaller and painted white, look rather like egg-cups (see Fig. 1.1). They sit on rails that reach across the platform.

There is nothing else to attract our attention. Not a soul in sight.

The Observatory

In 1986, I climbed this mountain for the first time, lost in the middle of one of the most barren deserts on Earth. At that time, the platform we see there today had not yet been built. The peak was just a kind of rocky ridge, almost without any vegetation at all, although I did pick up one rather unusual plant. With its woody stem, it long adorned my office at the Paris Observatory in Meudon. It could survive in these conditions of perfect drought only by absorbing the first dew to form on its leaves at dawn, before the burning sun would climb too high (see Fig. 1.2).

This peak was selected in 1983 by the Swedish astronomer Arne Ardeberg. It was such a promising site that it was made the subject of a systematic

Fig. 1.2   The mountain Cerro Paranal (2680 m), before the installation of the VLT, rises in the middle of nowhere, in the Chilean Atacama Desert. Credit: ESO/M.Sarazin

study. When I first arrived 3 years later, three Chileans—Francisco Gonzalez and his two sons Francisco and Italo—had been camping there for weeks on end to monitor optical and meteorological data. These would determine whether Cerro Paranal would be a suitable place to build the largest optical telescope in the world, a project proposed by a handful of European nations, including my own, France. Very dry weather conditions, a transparent sky, an extremely stable atmosphere, and very low levels of seismic activity were all part of the ruthless examination that Cerro Paranal would have to pass in competition with several other mountain peaks. When it was finally chosen in 1990, the top was levelled off by about 30 m using dynamite to make way for the construction of a platform that could accommodate the telescopes we hoped to build and which today constitute Europe's Very Large Telescope, the VLT, our wonderful telescope.

On this spring evening in 2018, more than 30 years after my first visit, I have returned to Paranal. That event in my youth is blurred and indistinct now in the foggy recesses of my memory. Meanwhile, night has fallen and myriad stars appear, incredibly bright, steady and untwinkling, so stable is the atmosphere, without a hint of turbulence. Their tiny specks of light combine to make the vast sky as bright as the broad expanse of the desert is dark, without village or lights as far as the eye can see. Lower down, a car suddenly breaks the silence of the desert and begins to make its way up the slope. It comes to a halt, parks, and three silhouettes emerge, only to be swallowed up a moment later by an entrance hollowed out of the mountain.

The scene changes now to this half-buried room where I meet the three silhouettes, in fact, three astronomers. It is from this control room that the eight telescopes on the platform are sent their instructions. The work stations are comfortable, each equipped with several screens covered with all kinds of measurement data. These pass on the commands that specify the night's schedule, already fed into the computers. If all goes well, those seated here will stick strictly to the plan, because every minute of observation counts. When the weather conditions are favourable, the night will always be too short to do everything one would have liked. Which way to point which telescope? Which instrument to choose this evening to analyse the light it gathers? Which measurement programme to carry out? For which team of astrophysicists, who will have made their proposal months before, then received acceptance and been allocated a certain number of nights of observation?

The reception tonight is a little out of the ordinary. To probe the depths of the Universe, only two teams will have access to the line-up of eight telescopes—the four giants with their 8.2 m mirrors, and the four others, measuring only 1.8 m, nestling in their white egg-cups. This is indeed unusual, because more often than not, there are half a dozen teams here sharing the instruments mounted on the telescopes, each running their own observation programme. The names of the four giants, whose birth will be recounted here, were suggested in the 1990s by a young schoolgirl, Jorssy Albanez Castilla, following a competition between schools in the Chilean II Region. The names put forward come from the language of the Mapuche people and refer respectively to the telescopes less poetically labelled UT1, UT2, UT3, and UT4, as Antu (the Sun), Kueyen (the Moon), Melipal (the Southern Cross), and Yepun (Venus).[3] We shall use these names here. The mirrors of these giants, each 8.2 m in diameter, place them among the largest optical instruments in the world, even in the present decade from 2010.

---

[3]The name 'Yepun' was originally taken to refer to the star Sirius, but later research on the language of the Mapuche showed that it actually meant the 'evening star' and referred to the planet Venus: https://www.hq.eso.org/public/teles-instr/paranal-observatory/vlt/vlt-names/yepun/.

A Distant Neighbour, the Black Hole in Sagittarius

The first team is led by Guy Perrin, an astronomer from the Paris Observatory. Tonight he is working with Frank Eisenhauer. Frank comes from one of the Max Planck Institutes in Garching, set in the middle of a superb science campus in the suburbs of Munich, in Baveria.

Guy was born in 1968 and became one of my keenest and most brilliant students, in those far-off days when he was studying physics and astrophysics at the *École polytechnique* and the University of Paris VII. During this account, we shall meet this eminent astronomer on several occasions, and we shall take the time to give a better introduction. This evening, his team has obtained permission to use the four giant telescopes of the VLT simultaneously and for several nights in a row. As surprising as it may seem, these four telescopes can work in concert, bringing together and mixing the light that each has collected. The aim of this observation programme, selected after tough competition between astronomers, is fascinating indeed. The idea is to obtain as precise and detailed an image as possible of what is going on in the immediate neighbourhood of a black hole, in fact, the supermassive black hole which almost certainly sits at the center of our galaxy, the Milky Way.

As astonishing as it may seem, the idea that an object as exotic as a black hole could exist in the Universe is not a new one, since it was first proposed in 1783. It was in that year that an English clergyman, John Michell, raised the possibility of a massive body, so massive that even light—whose speed was by then roughly established—would not be able to escape from its surface, due to the pull of gravity. Michell knew the mass of the Sun. He concluded that a star made from the same matter and of the same density as the Sun, but having five hundred times its radius, would be able to retain all the light emerging at its surface. This would be what he called a dark star.[4] Clearly though, such a star would be invisible to any telescope. But the Reverend Michell, in a stroke of genius, explained that one would nevertheless be able to detect its presence through the effects of its powerful gravitational field, in the ideal case where it formed a binary system with another, normal star.[5] By observing the latter, one should find an oscillatory motion that would reveal the presence of the

---

[4]On 27 November 1783, John Michell already had the idea of a black hole: www.aps.org/publications/apsnews/200911/upload/November-2009.pdf.

[5]"If the semi-diameter of a sphere of the same density as the Sun in the proportion of five hundred to one, and by supposing light to be attracted by the same force in proportion to its *vis inertiae* [inertial mass] with other bodies, all light emitted from such a body would be made to return towards it, by its own proper gravity."

invisible companion. Many double star systems had already been identified by that time, but nothing was known about how probable such a configuration might be. How likely was one to stumble across a binary object comprising a normal star and a dark star? It is not surprising therefore that there was such a long wait between Michell's suggestion and the date when the first black hole was discovered in 1973, through production of X-ray emission in its vicinity. This was the X-ray source subsequently referred to as Cygnus-X1. But it was a fruitful interlude and we shall return to it in Chap. 7.

Exactly in the center of our Galaxy—and here we use a capital letter, because this is the home of the Sun and its system of planets—, there sits an object some 26,000 light-years away with a mass about four million times the mass of the Sun. This is the only entity with such a high mass concentration in the Galaxy, and it has gradually become accepted that it is probably a black hole. We call it Sagittarius A*, because it sits in the beautiful constellation of Sagittarius, so glorious in the summer sky. The A* indicates that it is a source of light waves emitted at radio frequencies, and indeed, this is how it was discovered. Our German colleague Reinhard Genzel and his team began exploring the neighbourhood of this source in the 1990s. They launched the idea, which every further observation has since supported, that there is a supermassive black hole there, so close to us in space, just as it is in time for the light to travel to us—only 26,000 years, barely more than the time that separates us from the Neolithic. There are other black holes out there, but they are lost in space, a thousand times further away or more. This makes SgrA* a first rate laboratory, so much more accessible than all the others. From our vantage point on Earth, we can observe there some of the phenomena predicted by the general theory of relativity when the gravitational field becomes unbelievably strong. Sagittarius A* will therefore be one of the heros of our story.

This evening, the team seated at the control station will be coordinated by Frank from Bavaria, a colleague of Reinhard's, and Guy, two astrophysicists well-versed in their line of work. Each of the four operators, the only ones allowed to set their telescope in motion, will carry out the night's observing programme, passing on instructions to the corresponding computers. The latter will then obediently point the four telescopes Antu, Kueyen, Melipal, and Yepun toward the constellation of Sagittarius. They must follow the apparent motion of the stars in the sky, due to the steady rotation of the Earth on its axis during the night. Using carefully arranged mirrors, the light gathered by the four giant mirrors is then channeled to an underground room containing the impressive GRAVITY instrument built in the laboratories of Frank, Guy, and several others. Activated remotely by the control panels, this instrument can then receive, measure, and analyse the light which left the central region of the Galaxy some 26,000 years earlier, when Europe was still largely covered by ice. This light brings astronomers much crucial information. Carefully prepared computer programs are able to represent the results in the form of graphs, figures, curves, and numbers which will keep Frank and Guy busy throughout the night, along with half a dozen collaborators, scrutinizing the details of this complex celestial process (see Fig. 1.3).

The ultimate aim, doggedly pursued now for over a decade, is an extremely ambitious one. The idea is to distinguish each individual star within very close range of the black hole, held captive by the incredibly strong gravitational field it produces and hence orbiting around it at tremendous speeds. Tonight, the team are thus using the GRAVITY instrument to produce an unbelievably sharp image, sharper than any yet obtained in this region of the infrared light spectrum. Night after night, it is used to determine the changing position of one of these stars, called S2. The distance between S2 and SgrA*, as measured tonight, is only about fourteen light-hours. The sharpness of the image, which has to be good enough to distinguish the black hole from the star S2, is thus

Fig. 1.3   In the control room of the VLT and its interferometric mode (VLTI), French and German astronomers Guy Perrin and Odele Straub observe the center of the Galaxy with the GRAVITY instrument. Credit: Guy Perrin

simply characterised by an angle, namely, the angle between the direction of S2 and the direction of SgrA*, as viewed from the Earth. This angle is found as the ratio of two lengths, viz., fourteen light-hours divided by twenty-six light-years. This gives

$$\frac{14}{26,000 \times 365 \times 24} \sim 50 \times 10^{-9},$$

or fifty billionths of a radian,[6] which is about twelve thousandths of a second of arc, or twelve milliarcseconds. Put another way, the sharpness of the image obtained would be good enough to make out the silhouette of a rocket standing on the Moon! So this is the task undertaken by Guy, Frank, and the ninety-seven other researchers making up the GRAVITY team, the rest of whom will be eagerly awaiting news from the previous night, as they open their electronic mailboxes back in Europe the following morning. Because tonight, the measurements obtained with this unique instrument represent the culmination of a long and painstaking investigation which began 26 years ago. A few weeks later, a paper entitled "Detection of the gravitational redshift in the orbit of the star S2 near the Galactic center massive black hole" will inform the scientific community that, for the first time, one of Albert Einstein's most important predictions regarding the effects of a very strong gravitational field have been firmly supported.[7] Later, we shall see how they reached this point.

Exoplanets: Other Worlds So Close

In the vast room, silence reigns as everyone concentrates on the task at hand. For a second team, led this evening by Anne, is getting ready for the following night. Today, in the hours leading up to dusk, Anne has been carefully testing the instrument known as Sphere, set up at one of the focal points of the telescope Melipal. Her own laboratory contributed to its design and construction. When she has finished, she will hand over the use of Melipal to the GRAVITY team, but tomorrow night the telescope will be left to her, so she is doing a final check that Sphere will be able to carry out the intended programme (see Fig. 1.4).

---

[6]One radian is a unit of angle equal to the angle subtended at the center of a circle of unit radius by an arc of length π on the circumference of that circle. It is equal to $(180/\pi)$ degrees, or about 60°.

[7]This wonderful observation was published at the end of 2018 in the journal Astronomy & Astrophysics: Gravity Collaboration (2018).

Fig. 1.4   While the Sun is setting, Anne Lagrange relaxes before her observing night at Paranal. Credit: A. Lagrange

I first met Anne-Marie Lagrange—we shall call her Anne—when she was a young student at the *École polytechnique*, still trying to decide what path to follow in the world of science. Eventually, she opted for astrophysics and so it was that each week her keen gaze and determined face could be made out among the students following the master's lectures.[8] At the time, I was teaching at the University of Paris VII. Anne is another of the main characters in our story. After obtaining her doctorate between 1985 and 1988, when the European project was still in its infancy, Anne became interested in a star surrounded by dust, thought to have comets, like our

---

[8]What is now called a Master 2 in France was then a *Diplôme d'études approfondies* or DEA.

own star, the Sun. Now, whenever there is dust and comet formation, there may also be planet formation? The star β Pictoris and its ring are located in the southern constellation of Pictor. Like several other constellations in the southern hemisphere, this one was named after the voyages of Magellan, by the French astronomer Nicolas-Louis Lacaille around 1750. The star Beta Pictoris would become a springboard for Anne's amazing career. She would go from one discovery to the next, never losing her charm and simplicity, until in 2011 she was designated *Femme scientifique de l'année* in France.

Anne's research project is quite different from Guy's. Her quest is not to explore the mysterious and tumultuous environment of a distant black hole more than twenty thousand light-years from the Earth, but to prove the existence of a planet in orbit around another star only a few light-years away, and then establish its characteristics. Anne is an internationally recognised and respected expert on the subject of other worlds, or exoplanets. So why is this subject so important in the present story?

In Ancient Greece, Democritus invented the idea of the atom by pure intuition, more than 2000 years before Jean Perrin proved its existence and thereby obtained the Nobel Prize for Physics in 1926. Democritus was also interested in the existence of other worlds in the Universe. This was pure speculation, once again, but it is well worth quoting:[9]

> There are innumerable worlds of different sizes. In some there is neither a Sun nor Moon, in others they are larger than ours and others have more than one. These worlds are at irregular distances, more in one direction and less than another, and some are flourishing and some are declining. Here they come into being, there they die, and they are destroyed by collision with one another. Some of the worlds have no animal or plant life nor any water.

As the text makes so clear, these hypothetical other worlds are very different from the one associated with our own star. The analogies mentioned compare them with the Earth, but distinguish them by their diversity. Speculation about their existence has been going on for more than twenty-three centuries, continuing with *Conversations on the Plurality of Worlds* (1686) by Bernard Le Bouyer de Fontenelle, who was the permanent secretary of the Royal Academy of Sciences in Paris, or again the proposed existence of channels that the Italian astronomer Giovanni Schiarapelli thought he had found on Mars in 1877, and later still, the little green men I myself discovered in the comic strips of my youth.

---

[9]Democritus, lost text, quoted by Hippolytus in the third century AD.

At the beginning of the nineteenth century, the astronomer Pierre-Simon de Laplace suggested that the Solar System could have been formed from a cloud of gas, which he called the "primitive nebula". This meant that the Sun and planets actually shared a common history, and this mechanism could then apply elsewhere, not just to our own Solar System. Right through the twentieth century, Laplace's nebular hypothesis continued to guide research on what remained an enigma, the formation of the planets around our own star. Astronomers began to look more closely at nearby stars to see whether they too might have planetary systems whose existence could be proved by careful telescope observation. But unfortunately, the challenge was too great, because it was impossible to detect the faint light from these planets. They were just too far away to be able to make out on a photograph taken behind a telescope. In Chap. 6, we shall describe some of the stories that make up this century-long investigation, which remained fruitless.

After a series of somewhat hesitant announcements, there was a spectacular turn of events on 6 October 1995, when two Swiss astronomers at the Geneva Observatory, Michel Mayor and his student Didier Queloz, revealed the presence of a planet which would come to be known as 51 Peg b, with a similar mass to Jupiter. This planet orbits a nearby star rather like our own, 51 Pegasi, in the constellation Pegasus, visible to us in the northern hemisphere. Detection was indirect, since the planet is not actually visible on any image. What is detected are the gravitational effects of its presence on the motion of the star 51 Peg, revealed by a very high resolution spectrograph designed at the Marseilles Observatory by André Baranne. This discovery put an end to the suspense first sparked by Democritus and began a new era in astrophysics. As so often happens, the ground is prepared for a discovery, the question matures over a long period, and then one day the fruit falls from the tree. In 1992, the word 'exoplanet' was used for the first time. Today it refers to planets more or less like those in our own Solar System but orbiting other stars.

Thousands of exoplanets have been discovered in the Galaxy since 1995, and the current feeling among astronomers is that the majority of stars in our Galaxy probably possess one or more exoplanets, whence it is likely that the same will be true in the billions of other galaxies. This is an awe-inspiring prospect and indeed a major scientific revolution, providing a wonderful field of investigation for young scientists in the twenty-first century.

At the end of the day, the telescope Melipal, one of the four telescopes making up Europe's Very Large Telescope, is no more than a gigantic camera in which the objective is a mirror rather than a lens. A set of secondary then tertiary mirrors transmits the image formed at one of the focal points located to the side of the telescope. The millions of pixels of the light detector placed there, which uses precisely the same physical principles as the detectors in an ordinary camera, sends this digitised image to the computer screens in the control room. Anne has chosen the colour, that is, the wavelength of the infrared light she wishes to analyse, along with a host of other parameters that will be needed for her observations. The extraordinarily difficult challenge that Anne hopes to meet tomorrow night is to distinguish a tiny exoplanet which appears right next to its star and which is a million times fainter than that star, then form an image and analyse the light coming from it. Besides the need for an extremely sharp image, comparable to the sharpness obtained by the GRAVITY instrument, the problem is also to handle the radically different brightness of the star and its planet. This is more difficult than detecting a tiny mosquito flying around a distant lighthouse by the feeble light it scatters from the lighthouse lamp.

Thanks to her discoveries, Anne has become an internationally recognised specialist on exoplanets. The programme to be carried out by the Sphere instrument was designed by a team of a hundred and twenty-five scientists from all over the world, a group with which Anne is closely associated. The aim is to study a specific star called V1032 Centauri, or PDS 70, in the southern constellation Centaurus. This star is slightly cooler than the Sun and very young, being only about 5 million years old. It thus formed just a little before the birth of Lucy, the Australopithecus whose fossil was found in Ethiopia by the palaeontologist Yves Coppens and his colleagues. It has been known for about 15 years that this star is surrounded by an interesting disk of gas and dust which might have given rise to a planet. Could we then distinguish such a planet with absolute certainty, providing firm evidence for this relationship between disks of this kind and the formation of exoplanets? We shall discuss this relationship in more detail in Chap. 6. Thanks to the extremely sharp image produced by the Sphere instrument, the observations to be made over the coming nights will be combined with others acquired at Paranal over the last 4 years, and images from other telescopes, to try to confirm the suspicion that an exoplanet sits within this disk. This conclusion can then be further corroborated by observing the orbital motion of the planet in images taken at intervals of several months (see Fig. 1.5).

Anne tells me that this result would be the most beautiful among all those obtained by herself and her colleagues over the past decade.

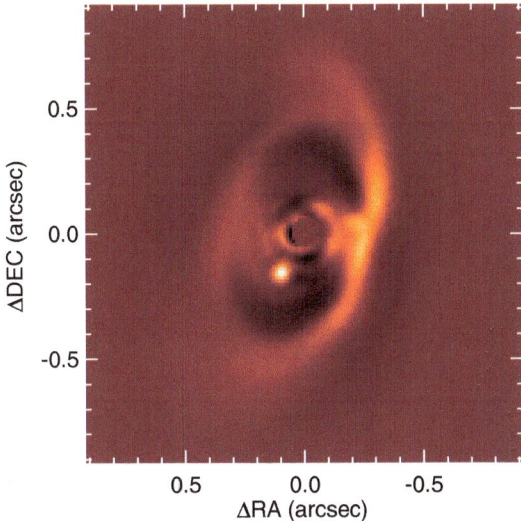

Fig. 1.5   The exoplanet PDS 70 b, discovered in 2018. In this small region (about one second of arc as seen from the Earth), with the instrument SPHERE on the VLT to suppress the blur, the planet shows up in the near infrared. Seen almost face on, the protoplanetary disc appears, with a dark zone where the material has gone to build the planet. The central star has been artificially masked to avoid overexposing the image. The orbit size is about 22 astronomical units (1 AU = distance from the Earth to the Sun). Figure from Müller et al. (2018)

Blurred Images

Using these huge telescopes, Frank, Guy, Anne, and the many astronomers making up their teams detect infrared light and obtain unbelievably sharp images, able to reveal the secrets of exoplanets and black holes. In the images produced by their instruments, displayed on their computer screens and analysed with considerable help from dedicated computer programs in their laboratories, they can make out the kind of detail that few would have even dreamt of 50 years ago. And when their discoveries are published, they are acclaimed the world over. In 40 years, and with the Very Large Telescope set up at the top of Cerro Paranal, Europe has succeeded in meeting a tremendous challenge, and been the first to get there.

So what exactly was this challenge? Fifty years ago, astronomical images, even those produced by the best telescopes of the day, equipped with large and perfectly polished mirrors, were afflicted by an inescapable blurring. What was the cause of this fuzziness, this lack of sharpness in the details? Was there something absolutely unavoidable that would forever limit what we could learn of distant objects when we tried to form their image?

This evening, at the top of Cerro Paranal, Guy and Anne have taken up the long and fascinating story of light which began several centuries ago, enacted by great scientists such as Huygens, Rømer, Newton, Young, Fresnel, Arago, Fizeau, Maxwell, Hertz, Planck, Einstein, and Feynman. Throughout this story, French scientists have played a major role, including several winners of the Nobel Prize in Physics, namely, Alfred Kastler (1966), Claude Cohen-Tannoudji (1997), Serge Haroche (2012), and Gérard Mourou (2018), all pioneers in the exploration of light.[10] This tale of lenses and mirrors takes us right up to the sophisticated instruments installed at Paranal. The astonishing discoveries made about black holes and exoplanets attest to the great successes of the past few decades in the relentless struggle to remove this blurring from our images. I chose to mention these discoveries at the very beginning of my account so that the reader would grasp the importance of this struggle. The black holes and exoplanets deserve more discussion, of course, and we shall return to them at length at the end of our story, once the victory over blurring has been secured.

So here, dear reader, in the following chapters, are a few episodes of this 50 year quest: the amazing story of our bid to beat the blur.

---

[10] Albert Fert, born in 1938 and a student of Jacques Friedel, obtained the Nobel Prize in Physics in 2007 for his work on solid state physics, another area where French physicists excel.

2

From the Dawn of Time

As for the future, your task is not to foresee it, but to enable it.—Antoine de Saint-Exupéry, *Citadelle*

The story of blurred images begins with the story of imaging, and this got under way 600 million years ago with the long history of the evolution of the eye in living creatures. It continued when the first lenses were made to improve people's vision, followed by the *camera obscura* or pinhole camera, used by Renaissance painters. Then came Galileo, who stood on the hills of Florence in 1609 and pointed the first astronomical telescope toward the sky. It was equipped with lenses and brought images to his eyes that no one had ever yet seen.

Before the invention of this refracting telescope with its lenses, it had long been known that, by simply making a small hole in the wall of a darkened room, the *camera obscura*, an image of the scenery outside could be formed on a screen inside the room. This extremely simple camera was described by Mo Zi in China in the fourth century BC, then by Aristotle. It was used by Leonardo da Vinci and Johannes Vermeer to study perspective. Maybe even by our ancestors in the neolithic. In the darkness of their cave, did they ever notice the rays of light entering by a small opening and forming an inverted image of the sunlit landscape on the wall at the back of the cave? As a distant echo of this idea, which I somewhat hesitantly put forward, I recall the pride of the physicist Yves Rocard during the explosion of the first French atomic bomb at Reggane in the Sahara on 13 February 1960. He amazed the high ranking officers seated around him in the bunker not far from the site of the

© Springer Nature Switzerland AG 2020
P. Léna, *Astronomy's Quest for Sharp Images*, Astronomers' Universe,
https://doi.org/10.1007/978-3-030-55811-6_2

explosion by using a small hole in the concrete wall to bring them, without risk, a spectacular image of the nuclear deflagration of *Gerboise bleue*. This story was passed around all his pupils at the *École normale supérieure* in Paris where I was studying at the time.

The image given by the *camera obscura* becomes less and less blurred when we reduce the size of the hole which allows the light to come in, as can be seen from a simple geometric argument. But the smaller the hole, the less light can enter and the darker the image. Painters got round this constraint by finding a suitable compromise between blurring and luminosity. With the advent of photography between 1816 and 1827, the *camera obscura* became obsolete. Two Frenchmen had the idea: Joseph Nicéphore Niepce and his later associate Louis Daguerre. It was the latter who reaped the glory when François Arago presented the discovery to the *Académie des sciences* in Paris on 8 January 1839 (Léna 2003). He showed an image of the Moon, which thus became the first heavenly body to be photographed. When it was made public, this new process for obtaining and reproducing images became known as the daguerreotype, thereby forgetting the scientific priority of Niepce, who died suddenly in 1833. History can be unfair. But so it was that photography began its own epic journey, and with it the struggle to eliminate blurring in the resulting images.

Astronomers quickly took hold of this novel technique, since they could now allow the light from faint objects to accumulate over long periods on a photographic plate, something the eye cannot do. There would be no going back! The days of observation with the naked eye were numbered. The telescope contains a lens, or more often than not a mirror today, to collect light and form an image at the focal point. The light distribution could then be fixed on a photographic plate using photochemical reactions releasing silver particles; or, for the last 50 years or so, using photoelectric detectors, sensitive to the different forms of light, from infrared to visible to ultraviolet, transforming them into electric currents. All the optical telescopes in the present account are variations on this simple setup. They are nothing but huge cameras, equipped with a photographic plate or an electronic detector to pick up and store an image with varying degrees of sharpness.

The Eye, Marvel of Evolution

Well before the advent of the *camera obscura* and the modern camera, nature had been exploring the problem of image formation in living beings. And she succeeded by inventing that extraordinary organ, the eye. This was some

600 million years ago, before the beginning of the geological era called the Cambrian. That was when the genesis of the eye got off the mark!

It is quite likely that at this time the evolutionary process discovered a light-sensitive molecule called proto-opsin, ancestor of all the photopigments which make the nervous system of an animal sensitive to the light in its vicinity. However, merely perceiving the presence of light, as might happen in a given cell, is a much more rudimentary task than forming an image with the help of an ensemble of cells, and the eye was still a long way off. At the beginning of the Cambrian, about 550 million years ago, a first eye would form at the surface of the skin of certain animals. It was a simple cavity, a tube working rather like a skylight, which allowed the animal to tell which direction the light was coming from. There can be no doubt that this primitive eye would give such a creature a genuine selective advantage. Indeed, fossils reveal the steady and rather fast improvement, until a third of the thirty-three branches of the animal kingdom at the time possessed some kind of eye. Six of those were able to form what could already be called an image. The organism in question would experience a set of simultaneous perceptions that could attribute an intensity and probably also a colour to each direction from which the light was received. Such an organism could then interpret this signal and adjust its behaviour accordingly.

The selective advantage gained by this ability was such that, with the remarkable diversity of their eyes, the modern day descendants of these six branches constitute 96% of all known animals, from the cephalopods to the amphibians and the mammals. Five hundred million years ago, the modern visual system was already present with all its key components among the vertebrates that would give rise much later to humans. One of these components is the iris. This forms a diaphragm called the pupil, whose aperture varies automatically depending on the light intensity. Then there is the lens. This forms an image on the retina, which is itself just a layer of photosensitive cells, and it also focuses the light correctly depending on the distance to the light source. And finally, there is the visual nervous system, which processes the information received and supplies the living creature with a perception.

Later in this story, we will often come across a term that was first heard among astronomers only during the 1980s: adaptive optics. And yet, evolution had already invented it in the natural world! The primitive visual system which came into existence 500 million years ago already had several of the feedback mechanisms required. If there is too much light, at levels that would swamp the retina, the muscle of the iris shrinks the pupil. In darkness, the opposite happens and the pupil dilates. Moreover, depending on the distance of the object to be viewed, a muscle controls the lens of the eye, which has a certain elasticity, curving the surface to differing degrees and thereby changing its focal point to produce a clear image. The direction of the animal's gaze can also be adjusted by rotating the eyeball in order to use the best part of the retina and the photoreceptor cells that cover it. Despite these wonders, which have raised many questions regarding their evolution, we should not forget certain design failures. For example, when the cephalopods separated from the vertebrates in the late Cambrian period, the eyes of the vertebrates evolved in such a way that the blood vessels irrigating the retina were placed in front of the photoreceptors, forcing the animal's brain to correct an image that is seriously marred by the shadow of these vessels. The octopus does not have this problem because the blood vessels pass behind the photoreceptors.

Our own retina, result of this evolution, can achieve quite extraordinary feats that photographic plates have never been able to match. The human eye, when it gets accustomed to the darkness, is almost capable of detecting individual photons one by one when they have wavelengths lying in the visible range of the spectrum, between blue and red. Such levels of sensitivity only became possible with the advent of modern electronic sensors in the form of charge coupled devices (CCD), of the kind that now equip ordinary cameras.

Notwithstanding, the human eye cannot supply our brains with perfect images, because its visual acuity, that is, its ability to make out very fine details, is actually rather limited. There is no problem making out the Moon's disk at night, because the apparent diameter of the Moon subtends an angle of half a degree (thirty minutes of arc): one would have to juxtapose 180 such disks to go from the horizon to the zenith, corresponding to 90°. We can also discern some smaller details on the Moon's surface, such as the *maria* or seas, and perhaps just about distinguish a little crater subtending one minute of arc.

When a detail subtends an angle less than one minute of arc, it becomes blurred in the retinal image and the detail will be unrecognisable. Whatever is done to try to remove this blurring and make out smaller details in the landscape, the task is hopeless. The eye has reached its limits. Even better than the lunar surface as a test of visual acuity is a beautiful cluster of a dozen or so bright stars called the Pleiades, which can be seen in the winter in the constellation Taurus. For many observers, it remains blurred and they are unable to count the stars. But children, who often have better visual acuity, are happy to take up the challenge, and some can count up to eight stars in this tiny cluster. Our ancestors, too, because on the rocks of Mont Bégo, in the Vallée des Merveilles of the Mercantour massif in southern France, there are thousands of drawings. Many show cup and ring shapes carved in the sandstone some 5000 years ago. As explained by the palaeontologist Henry de Lumley, one of these drawings corresponds exactly to the arrangement of the stars in the Pleiades. As noted by the priests, when it rose above the horizon just before the Sun, it announced the sowing season.[1] This is reminiscent of the heliacal rising of the star Sirius a month after the summer solstice, which announced the flooding of the Nile in Egypt, so important for their agricultural cycle.

Some people who have exceptionally good eyesight can make out details measuring less than half a minute of arc, as can birds of prey like eagles and falcons. Legend has it that the lynx has such good vision, but we now know that they are shortsighted. However, the idea was used to name the *Accademia dei Lincei*, the Italian science academy, because such sharp vision was supposed to symbolize the observational prowess required by science. Assuming the human eye to be in good condition, unaffected by short-sightedness, long-sightedness, or astigmatism, if it were only limited by the fundamental properties of light itself, to which we shall return in the next chapter, its resolution would be three times as good and hence comparable to the falcon's. The astronomical telescope and the microscope were invented to make up for the short-fallings of evolution, which did not in fact endow the human being with such performance, considerably reducing the blur in the images available to us.

Reading Glasses and Telescopes

As we grow older, our visual acuity and accommodation can deteriorate significantly. Hence the invention of the eyeglass, then reading glasses. Such man-made instruments were the first artifacts to be able to reduce the blurring of images formed on the retina. Reading stones were made of rock crystal (quartz) in the Middle Ages to help monks copying documents in monastery libraries. In the thirteenth century, with increasing understanding of the paths light takes in different media, and in particular of the effects of refraction, obtained by scholars like Al Aitham and Roger Bacon, these lenses were improved by using glass, giving rise to the first pairs of spectacles that could sit on one's nose. But it would be many years before the advent of the refracting

---

[1]Annie Echassoux, Henry De Lumley, Jean-Claude Pecker, Patrick Rocher, *Les gravures rupestres des Pléiades de la montagne sacrée du Bego, Tende, Alpes-Maritimes, France.* https://halshs.archives-ouvertes.fr/hal-00480210/. The Nebra Sky Disk, discovered in Germany in 1999 and dated to around 1600 BC, is so far the oldest extant representation of the celestial sphere, including the Pleiades. The rock carvings in the Vallée des Merveilles may well turn out to be older. https://en.wikipedia.org/wiki/Nebra_sky_disk.

telescope as used by Galileo. The latter was not its inventor, and indeed there is some doubt about who actually was. It may have been the Dutchman Jacques Metius who exported the idea to Italy around 1608, or another Dutchman, Jacques Lippershey. However, it was Galileo Galilei who, at the age of 44, pointed just such a telescope toward the island of Murano on 21 August 1609 to demonstrate its virtues to a Doge who would certainly have been interested in its military applications for the benefit of the Most Serene Republic of Venice. Then, over the following months, this same Galileo turned it toward the sky, the Sun, and the stars. It was thus that he saw images of the heavenly bodies with a detail unknown to any before since the dawn of humanity. One can only imagine the excitement that must have filled him, as expressed so well by Bertolt Brecht in his *Life of Galileo*: "And the Earth rolls joyfully round the Sun, and the fishwives, merchants, princes, and cardinals, and even the Pope, roll with it."

So what exactly made this telescope better than the human eye? To begin with, compared with the pupil alone, it could gather a hundred times more light in the eye of Galileo and the eyes of the people of Florence who were able to try it, simply due to its diameter of several centimetres. This meant that it could observe faint objects, such as the four natural satellites of Jupiter, which Galileo discovered and called the *Medicea Sidera* or Medician stars, in honour of his Florentine patrons. But above all, thanks to the diameter of the lens, the telescope improved by a factor of ten on the famous limiting angle of one minute of arc, due to the imperfection of the eye and responsible for the lack of sharpness in the details when they are observed with the naked eye. While Galileo did indeed discover that Saturn had a strange appearance when observed with his telescope, the image remained too blurred to identify the presence of its rings. Instead, he interpreted the elongation of the image, revealed by his telescope, as being due to a triple planet, i.e., consisting of three bodies. It was not until 1665 that the Dutchman Christiaan Huygens, using a telescope of better optical quality, was able to correctly describe "a thin, flat ring, nowhere touching, and inclined to the ecliptic" (see Fig. 2.1).

From the seventeenth to the twentieth century, the techniques for producing, polishing, and testing glass were steadily improved, and it became possible to build instruments with ever larger aperture, hence capable of gathering more and more light from the observed bodies. After the refracting telescopes, which are instruments equipped with lenses, there came the reflecting telescopes, equipped with mirrors, like the four giants in Paranal. The optical quality of these instruments got better all the time, allowing us to go beyond the limited acuity of the naked eye and form ever more precise images. Let us see how this was done.

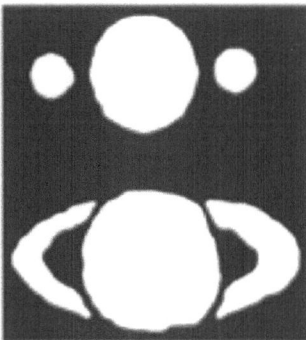

Fig. 2.1 Original drawings of the planet Saturn, observed by Galileo in 1610 (*top*) and in 1616. The image blur did not allow him to distinguish the ring. He interpreted the lateral spots as companion bodies but was puzzled by their non-circular shape. Changes of appearance between the two dates are due to the changing inclination of Saturn's rings, as seen from Earth

Atmospheric Disturbances

In the race to discern ever finer detail in astronomical imagery, a new limit would eventually appear, and it would not be the last. The great Isaac Newton, a complex character in the world of science if ever there was one,[2] may not have been the first to encounter it, but would certainly be the first to give a correct description. He invented the telescope that is now named after him and he polished the mirrors himself. He was aiming for the ultimate quality in his images, whether of the Moon, the planets, or the stars. One of his aims was to separate two stars so close together that they form a double star, a system that a less powerful instrument would only show as a blurred spot. When he observed this with his telescope, Newton found that the greatly enlarged images formed a continually distorting spot, and that increasing the magnification of the telescope merely enlarged this blurred spot for the eye of the observer, while it remained just as blurred as ever. He understood that this incessant agitation of the image was due to the thickness of air in the atmosphere, itself agitated, which the light from the star had to cross to reach the telescope. It is precisely this agitation which, by mixing layers of air of different temperatures, causes the stars to twinkle so poetically, in the phenomenon known to astronomers as scintillation. This is something anyone can observe, especially when the

---

[2]There is no shortage of books about Newton, including the excellent biography by Westfall (1994). The mysterious complexity of this person has been analysed by Loup Verlet in *La malle de Newton* (Gallimard, 1993).

weather is about to change, since then cold and hot air mix together. These non-uniformities disturb the path of the light through the air.

With remarkable foresight regarding the best places to locate observatories, Newton made the following recommendations in his famous treatise *Opticks* (1704):

> For the Air through which we look upon the Stars, is in a perpetual Tremor [...]
> Long Telescopes may cause Objects to appear brighter and larger than short ones
> can do, but they cannot be so formed as to take away that confusion of the Rays
> which arises from the Tremors of the Atmosphere. The only Remedy is a most
> serene and quiet Air, such as may perhaps be found on the tops of the highest
> Mountains above the grosser Clouds.

And here is his description of a star when observed through a telescope:

> [...] one broad lucid Point, composed of those many trembling Points confusedly
> and insensibly mixed with one another by very short and swift Tremors, and
> thereby cause the Star to appear broader than it is, and without any trembling
> of the whole.

This is indeed, the earliest and most precise definition of blurring. Something that had to be understood, then overcome.

Following Newton's remarks and with whatever means of transport were then available, astronomers sought high mountains for their observatories, where the atmosphere was as calm as possible, to reduce this agitation so harmful to the quality of their astronomical images. However, the blurring would resist for more than three centuries, and was only finally overcome in the middle of the twentieth century, as we shall discuss in more detail below. Astronomers use the word 'seeing' to speak of this. The site of any given observatory is characterised by its seeing conditions. The seeing is measured by the mean angular size of the seeing disk obtained when the telescope forms an image of a star. The angular size of the seeing disk is very small, of the order of a second of arc.

The seeing disk is so important in the present story that it is important to be particularly clear about it. If we take a look at the beautiful blue star Rigel, in the constellation of Orion, without using any kind of telescope, it appears as a single point of light, given the limited resolution of the human eye. However, with a modest telescope, we can make out a double star: two points each spread over about one second of arc by the seeing effect, but standing about ten seconds of arc apart. In this case, the seeing effect can no longer conceal the presence of two companion stars. On the other hand, if one of these two companions were itself a double star with a separation of a few hundredths of a second of arc, we would not be able to make them out individually. They would be lost within a single seeing disk.

At Cerro Paranal, the Chilean mountain where we began this story, the seeing is slightly better than one second of arc, although better still on exceptionally calm nights. The second of arc is a useful benchmark when discussing the blur in images. While 1 min of arc represents 115 km on the Moon when viewed from the Earth, one second of arc represents 1/60 of that, or about 1900 m.

After Newton, astronomers began to build refracting telescopes, then reflecting telescopes of ever greater diameter—first decimeters, then meters. By gathering more and more light energy, due to their larger diameters, these telescopes can form images of fainter and fainter objects. In Newton's words, they "cause Objects to appear brighter" and hence detectable. But increasing their magnification cannot eliminate the "confusion of the Rays" in the blurred disk of the image. For several centuries, this atmospheric limitation due to seeing meant that the acuity of astronomical images could never be much less than a second of arc. But this was nevertheless sixty times better than the sharpest human eye.

The Shadow of a Hair

Since the atmosphere imposed a barrier on the acuity of any telescope observing the stars, and one that was long considered insuperable,[3] it was natural to wonder what could be done with an instrument gathering light that had not first passed through an agitated medium; for example, a telescope placed in space, outside the Earth's atmosphere. Would such an instrument be able to obtain much greater, even perfect acuity? Once again, the story is worth telling, because this is the story of light itself, in all its mystery.

In France in 1648, Louis XIV was just 10 years old, and trouble was stirring among the nobility in the form of the *Fronde*. It was around this time that Blaise Pascal, who had learnt about the experiments by Florentin Torricelli,

---

[3]As we have seen, the term 'visual acuity' refers to the ability of the human eye to make out fine detail. There exist various measures for acuity. Using the foot as unit of measurement, the acuity is expressed relative to 20/20, this being the best value, although some people with exceptional eyesight can have a slightly higher value. In this book, by a suggestive and practical analogy, we shall also use the word 'acuity' to characterise telescopes, measuring their ability to form detailed photographic images. The technical term is 'angular resolution', which will be introduced shortly.

demonstrated that it was indeed possible to obtain a vacuum, publishing his *Expériences nouvelles touchant le vide* (1647). Then, a year later, on Pascal's recommendations, his brother-in-law Florin Périer climbed to the top of the Puy-de-Dôme with a mercury column and the results were published in *Traités de l'équilibre des liqueurs et de la pesanteur de la masse de l'air* (1663).[4] These were great moments for physics, which was by then greatly concerned with experimental validation, as Galileo had advocated a few decades earlier. After the vacuum and the air, the next subject to stimulate the curiosity of these seventeenth century scientists, keen to understand everything that nature could offer, was light.

Francesco Maria Grimaldi was a Jesuit priest, philosopher, and teacher. He was particularly interested in mathematics and physics in the then flourishing town of Bologna, where the oldest European university had been founded in 1088, and which had been depicted in the previous century by the painters of the Carracci family. Grimaldi got to hear about a curious phenomenon that had been observed and reported. When a very fine hair is illuminated by an intense but small source of light, in such a way that its shadow is cast on a white screen, this shadow will be sandwiched between thin dark fringes that lie parallel to the hair and between which one can make out the colours of the rainbow (see Fig. 2.2). So what is the hair doing to the light to make its shadow behave so strangely? Like any good modern scientist who believes in experiment more than in his own preconceived ideas, Grimaldi, then 30 years old, carried out the experiment under a range of different conditions. He observed the same colours as those of the rainbow and coined a word to refer to this phenomenon, calling it the diffraction of light. This term actually combines two different phenomena that would be isolated in the seventeenth century to characterise different kinds of behaviour of light: dispersion, studied by Newton when he decomposed white light from the Sun using a prism, and refraction, studied by Descartes, who investigated the change in direction of light when it passes from air into water. But Grimaldi did not stop there! On the basis of his observations, he also suggested that light was a fluid with a rapid wavelike motion, thereby introducing an idea that he could not really demonstrate himself, but which turned out to have a good future.

Indeed, a 150 years later, at the beginning of the nineteenth century, there was considerable debate in Europe about the fundamental nature of light. On the one side were the supporters of Newton, for whom light was a flux

---

[4]The mercury column was shorter at the top of the mountain, which showed that atmospheric pressure diminishes with altitude, as Pascal had predicted (Pascal 1957).

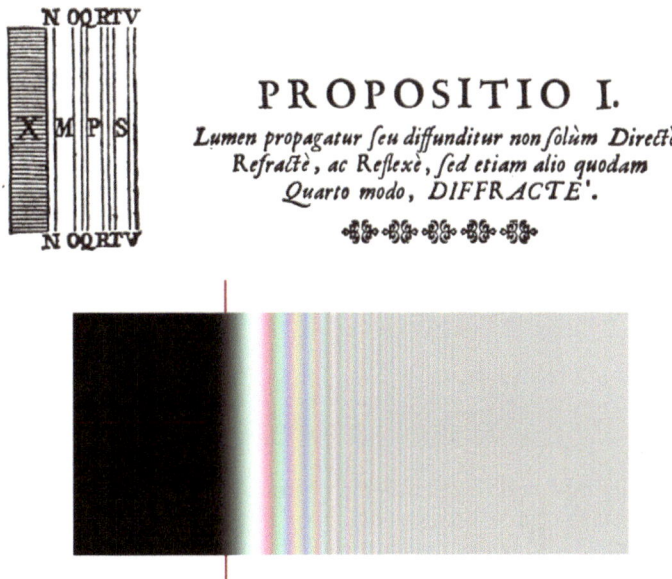

Fig. 2.2 In his book (1665), Francesco Grimaldi reports the observation of colored bands in the shadow of an illuminated straight edge (*upper left*), and coins the term 'diffraction' to describe this phenomenon (*upper right*), shown on a modern image in white light where fringes of different colors show different spacings (*lower*). Credit: Upper figures: Grimaldi (1665). Credit: Lower figure https://melusine.eu.org/syracuse/mluque/fresnel/augustin/diffractionbord.html

of tiny corpuscles, and on the other, those of the British scholar Thomas Young and the French engineer Augustin Fresnel, for whom light was a wave, analogous in some ways to the acoustic waves that transport sound through the air. The latter were gradually gaining ground, demonstrating new phenomena that supported their view, such as Young's famous fringes. Apart from this, they were developing mathematical tools that could be used to model these phenomena to great accuracy.

Here is one of the new phenomena that was particularly intriguing. A beam of light passes through a small circular hole, then encounters a carefully polished converging lens. This lens forms the image of the hole on a screen, whence the image is a small disk of light. As the size of the hole is decreased, the diameter of this small disk also diminishes. The surprise is that, below a certain diameter of the hole, the image disk no longer gets smaller but reaches a limit. Furthermore, rings of rainbow colours appear around it! This new phenomenon is attributed to diffraction, which could now be explained and calculated by the wave theory of light, as formulated by Fresnel. As

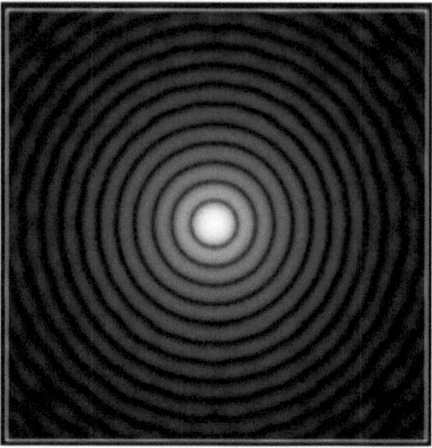

Fig. 2.3 The Airy disk and rings. A perfect telescope, with no atmosphere above, observing a pointlike source, provides an image showing a blurred disc, surrounded by luminous rings. Credit: Eric Gendron

with the hair, light spreads out when it encounters an obstacle. The disk surrounded with rings was studied in detail by a Cambridge professor by the name of George Biddell Airy. In 1835, he identified exactly what obstacle was responsible for this effect, in fact, the edges of the lens, and he determined the value of the limit reached by the image. The name of Airy would subsequently be immortalised by the term 'Airy disk', which has since become a *pons asinorum* for first year optics students (see Fig. 2.3). To a good approximation and up to a factor close to unity, the angular size of the disk is given by the ratio of the light wavelength and the diameter of the lens (or mirror) used to form the image. So with yellow light and a lens of diameter 10 cm, the disk will have an angular size of one second of arc.[5] It will be ten times smaller if the lens measures 1 m in diameter, and a hundred times smaller for a mirror of diameter 10 m, which is almost the size of the mirrors installed in the four giants at Paranal.

As a curious child, I loved to visit the wonderful *Palais de la Découverte* in Paris, originally inspired by the physicist Jean Perrin during the years of the *Front populaire*, an alliance of left wing movements formed in France in 1936. The idea was to make science accessible to everyone. He wanted to show

---

[5]This angle can be transformed into a length, namely, the size of the Airy disk on a photographic image placed at the focal point of the lens. If the latter has a diameter of 10 cm and a focal length of 1 m, the Airy disk with its coloured rings will measure 5 μm in the image and a magnifying glass would be needed to see it clearly.

how beautiful certain experiments could be, especially those which, in his own words, "replace what is visible but complicated by what is invisible but simple". I particularly liked the darkened rooms of the optics experiments, where one could play colourful games with light, producing effects like Grimaldi's hair or the Airy disk, which were quite mysterious to me. And when these fringes and rings were presented at school and explained by our physics teacher, Monsieur Yves Laurent—we admired him because he had joined the Royal Air Force as a pilot during the Battle of Britain—, I reproduced them myself many times and they left their mark in my mind. As a light source in our apartment, much to my mother's dismay, I put together an electric arc using graphite rods removed from the batteries of a torch (the Wonder battery only wore out if it was used, according to the advertisement!). Photographic paper was perfect for the screen, and I developed it with great anticipation, because the interference fringes caused by diffraction were indeed recorded there.[6] I did not realise at the time that a good part of my professional life would be taken up by the study of these same fringes and a relentless struggle to reduce the blurring in images!

So let us return to telescopes and the details in their images. Even in the twentieth century when a large telescope was launched into the vacuum of interplanetary space, where there is no air and hence no problem of seeing, this instrument would still not be able to make out the very finest details. It still could not achieve what I would call, by analogy with the eye, a perfect acuity. For example, when photographing a star, the image would still appear as a diffraction disk, a small blurred disk surrounded by Airy's rings, that could never be made smaller than the limit determined by the diameter of the telescope. If the star is double, and if the two components are closer than the size of the disk, the photograph will not be able to distinguish them. It is the nature of light itself that is the root cause of this blurring. If an instrument is placed in space, it will do better than the limit imposed by the seeing conditions on an Earth-based telescope. In this way, we can reduce the blurred disk, but without ever being able to completely remove it. This diffraction limit, which affects all images, can only be reduced by increasing the diameter of the mirror.

In 1990, NASA associated with the European Space Agency (ESA) to launch the large telescope called the Hubble Space Telescope (HST) into orbit around the Earth at an altitude of 600 km. Its main mirror measured 2.4 m

---

[6]This may have been the remote beginnings of our hands-on project *La main à la pâte*, launched in 1995 to introduce primary school children to the pleasures of science. See Charpak et al. (2005). See also https://www.fondation-lamap.org.

in diameter. The US military could mass produce such mirrors, of excellent quality, for their spy satellites in low Earth orbit. While a 1 m mirror would give a diffraction disk of 0.1 s of arc for yellow light, the Hubble mirror, limited solely by diffraction, should have had a disk measuring only 0.04 s of arc, roughly twenty times as good as the best seeing that could be obtained by an Earth-based telescope, and without the slightest risk of being blocked by a cloud! But astronomers would soon be disappointed because, just after Hubble was launched in 1990, it was discovered to be disastrously short-sighted. Its optical performance was marred by a polishing error that had escaped notice despite the multitude of tests carried out by NASA throughout the instrument's long development. And so it was that the acuity of HST was no better than that of an Earth-based telescope.

It became essential to fix the problem, and 3 years of studies were needed to come up with an optical solution to correct the defect. This was built, then carried to the orbiting telescope by mission STS-61 of the space shuttle Endeavour. There were three astronauts aboard, responsible for making the repair. One of these was my Swiss friend Claude Nicollier. Ten years earlier while he was still in training for the space flight, I flew over California with him aboard NASA's four-engined Convair-990, and then from Europe in the elegant Caravelle jet aircraft, in a bid to detect infrared radiation from the stars before it could be blocked by the water vapor present in the low Earth atmosphere. The space shuttle perfectly succeeded in its rendez-vous with the HST. A mechanical arm operated by Claude was used to grab the telescope and hold it firmly alongside the shuttle, which was moving at close to 30,000 km/h relative to the Earth at the time. The other two astronauts put on their spacesuits and went out into space to set up the instrument that had been designed to correct Hubble's short-sightedness. The mission STS-61 was a total success, and HST went on to provide magnificent images whose details are only limited by diffraction, leading to an exceptional harvest of discoveries. The end of its mission is planned for 2020, after 30 years of loyal service.

Determination in the Face of Adversity

The reader should be under no illusion that the insurmountable barrier posed by the Earth's atmosphere for the acuity of telescopes built on mountains prevented further astronomical discovery through the nineteenth century and the first half of the twentieth! Astronomers just had to accept this limitation imposed by nature and went ahead with determination in the face of adversity. The chemical composition of stars, the sources of the light they emit, the nature of interstellar clouds, the expansion of the Universe, and cataloguing of the multitude of other galaxies are just a few of the advances made thanks to improvement of Earth-based observatories, and this despite the fateful blurring which by then seemed inescapable unless space travel could one day provide a way around it.

On maps of the Earth's surface, a latitude and a longitude are attributed to each point. These are of course the coordinates given by the GPS. Perhaps, from time to time, the reader is curious enough to consult sky charts and catalogues, where the positions of the stars are also specified relative to certain coordinate systems. He may be surprised to find that, from the beginning of the twentieth century, the positions of the stars in the old star catalogues are given with an accuracy more than ten times better—a tenth of a second of arc—than the seeing limit. The reason is that a series of many successive and repeated position measurements yielding blurred and agitated images can nevertheless be used to determine a better astrometric position of the star by averaging. Indeed, a positional accuracy ten times better than the seeing limit can be achieved in this way.

For the moment, there were no black holes or exoplanets among these discoveries. Astronomy would have to wait a little longer for those. Despite the beauty of all these discoveries up to the middle of the twentieth century, the seeing limit made it impossible to image the surface of any other star than the Sun, and even an image of the latter made using an Earth-based telescope would have no fine detail.[7] So did this mean that there was no hope of ever imaging the disk of a bright star like Betelgeuse or Vega, and obtaining an image comparable to those of the Sun, in which we can make out certain details, just as a telescope can photograph sunspots? By convenience I suppose, or perhaps in despair, my teachers told me, and the textbooks confirmed, that a star would forever remain "like a geometrical point of no spatial extent" for us humans. This was a truly unbearable situation.

But then, toward the end of the nineteenth century, two physicists opened a tiny window of hope. One of them, Hippolyte Fizeau, was French, and the other, Albert Michelson, was American. What they did was to devise a subtle method that would play a key role in our story, because from 1970 it gave rise to an exceptional series of developments that eventually led to the little 'egg-cups'

---

[7]The Paris Observatory has some beautiful photos taken at the observatory in Meudon at the turn of the nineteenth century by the astronomer Jules Janssen. They show a fine structure at the surface of the Sun, known as granulation, but these images are nevertheless affected by the same limitations on their acuity and the details of the granules that we can now make out by observation from space just cannot be discerned.

and great giants that we have already met at the top of Paranal. This method, as ingenious as it was difficult to implement, had been used to measure the diameters of a few nearby stars at the beginning of the twentieth century, but had then been more or less left to one side as a kind of laboratory curiosity. In any case, thanks to this idea, at least some stars were no longer just geometrical points. We shall return to this in Chap. 4, because it will cast light on the rest of our story.

3

Too Good to Be True? Adaptive Optics

Mirrors would do well to reflect a little more before sending back images.—Jean Cocteau, *Blood of a Poet*

Impossible is not French, my father used to say to me, rather to encourage me to explore the unknown than in a spirit of extreme patriotism. And indeed, it was a young Frenchman, Antoine Labeyrie, who in the early 1970s was the first to see a way to break through the seeing barrier which had so long frustrated the astronomer's curiosity. Over the next 20 years, the breach was then steadily widened. These were precisely the years in which Europe, stunned by two world wars but confident in its potential, the result of centuries of exceptional scientific contributions, was looking for ways to make itself a new future through ambitious joint projects. One of these was the construction of an optical telescope, that is, operating at light wavelengths between the blue and near or mid-infrared,[1] that would become the most powerful in the world. Powerful in every sense of the term, because it would gather more light than all its predecessors then in use in the United States or the Soviet Union, but also because it would make a serious attempt to improve on their acuity.

---

[1]In the following, we shall use the term 'visible light' to refer to light whose wavelength lies between that of violet and that of red light, i.e., in the range 0.35–0.70 μm. The terms 'infrared' or 'near infrared' refer to light with wavelengths between 0.7 and 6 μm, while 'mid-infrared' corresponds to wavelengths up to about 12 μm. The Earth's atmosphere is of course transparent to visible light, and also to near and mid-infrared, but it is practically opaque beyond that. The somewhat vague term 'optical region' covers more or less all these wavelengths.

© Springer Nature Switzerland AG 2020
P. Léna, *Astronomy's Quest for Sharp Images*, Astronomers' Universe,
https://doi.org/10.1007/978-3-030-55811-6_3

Breaking Through the Seeing Barrier

Antoine often fascinates, but sometimes annoys those who meet him. In my view, he perfectly exemplifies a long tradition of excellence in French optics, without which none of what I am about to recount would actually have been possible. His father, Jacques Labeyrie, was a scientist and assistant of Frédéric Joliot-Curie at the *Collège de France*. He himself has boundless admiration for the inventiveness of nature, and has often tried to copy the ingenious technical solutions discovered by evolution, like legs and wings. He is an ecologist, both scientifically and in his way of life, leaving his solar house in Provence to go to his laboratory on his Merens mountain horse, an endangered species from the Ariège department of France. Sometimes deliberatively provocative, in his youth he once declared that he could have built the Hubble Space Telescope in his garage for a budget a hundred times less. A champion of participatory science, when he was appointed professor at the *Collège de France*, he brought together a whole range of enthusiasts, not just 'official' research scientists. We shall meet Antoine on many further occasions in the following pages.

In 1970, at the age of twenty-six, six years younger than myself, he left the *École supérieure d'optique* in Paris as a qualified engineer and presented his doctoral thesis to the jury at the Sorbonne. The core section of this work, prepared at the Paris Observatory in Meudon, where he shared an office with me, could be summed up in three pages. But what pages they were! So what was it all about?

A Bright Idea

Looking closely at the image of a star, affected as usual by the fuzzy seeing disk, several particularly attentive observers had noticed a curious phenomenon. Far from being uniform, the disk exhibited some very fine detail in constant agitation, although too fine and fast-moving to be made out properly with the naked eye. Clearly, this was lost when a photo was taken because exposure times of a few seconds would smooth out such details and produce a uniform disk of light. Back in 1958, the director of the Pic-du-Midi Observatory in the Pyrenees, Jean Rösch, had even given a name to this phenomenon. He described the image as being like grape seeds, or rather a bunch of grapes. The structures he observed contained details that could be clearly made out. A few years later, a French optical engineer at the Paris Observatory, Jean Texereau, gave a careful description of the phenomenon and interpreted it in the same way as had Newton, attributing the fine details present in the seeing disk to the rapid agitation of the Earth's atmosphere that the light had to cross on its way from the star.

Antoine went further, because he understood the physical origins of these fluctuating details. He found the exact relation between the size of the grains in the grainy structure and the size of the Airy disk produced by the primary mirror of the given telescope. This was a crucial step forward. While it had always been thought that the seeing would wipe out any finer details, it turned out that an instantaneous seeing disk actually contained much tinier ones. Indeed, their size is precisely the size of the Airy disk specifying the diffraction limit. So not all the information about the finer details of the star is actually lost when the light crosses the Earth's atmosphere. Some information survives, although distorted, and this is what produces the appearance of grape seeds (see Fig. 3.1). The same grainy effect, known as speckle, is produced by the spot of light on a piece of paper illuminated by a laser. Like every student in those days, in admiration of the new invention, Antoine had manipulated such lasers himself and had made the connection. As often happens in science, an unexpected but clever connection can suddenly throw light on a problem. As Louis Pasteur once said, "fortune only favours the prepared mind".

In his short and lucid paper, Antoine set out a method for recovering this information contained in the speckles (Labeyrie 1970). After paying homage to the precursors of the idea in the last century, Fizeau and Michelson, as mentioned above, he thus showed that a detailed image of the star surface could be rather simply extracted. When I spoke about the content of Antoine's paper, I said that it constituted a crucial step forward. Indeed, since 1970, this

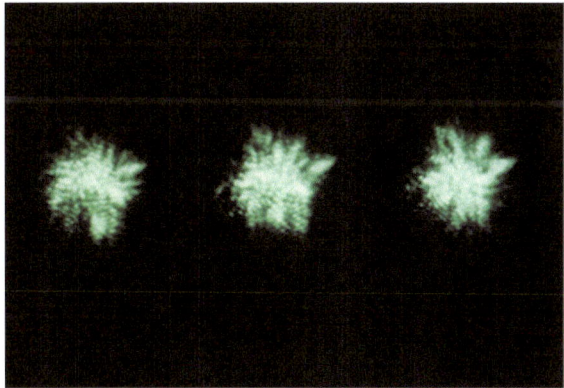

Fig. 3.1 Successive short exposures of the image of a 'point-like' star through a large telescope. The seeing disk changes constantly. It contains much smaller structures whose size is fixed by the diffraction limit of the telescope. Credit: G. Weigelt

paper has inspired hundreds of scientists in their quest to improve the detail in astronomical imaging.

Antoine has ever since been putting his ideas to the test, naturally on a telescope with as large an aperture as possible. At the time, the most powerful optical telescope in the world was in California, on Mount Palomar near Los Angeles. It was named after its instigator and designer, George Ellery Hale, a visionary American astronomer who, in 1926, had pushed for the construction of a mirror twice the size of the previous largest instrument, the Hooker Telescope, inaugurated in 1917, with which Hubble had discovered the expansion of the Universe in 1929. The Hale Telescope would thus measure 5 m in diameter—or more exactly, 200 in., as the United States had not yet adopted the metric system. It was financed by Rockefeller and associated with the California Institute of Technology (Caltech), a private university. Its first image was of a gas cloud in the Galaxy, taken in January 1949. It was the famous astronomer Edwin Hubble who took this photography—and it was in homage to his discoveries that the Hubble Space Telescope (HST) got its name. The Hale Telescope remained the largest optical telescope in the world until 1976, when the 6 m Soviet Bolshoi Telescope became operational in Zelenchuk, in the Greater Caucasus.

So 1 year after his doctorate, in 1971, the young Antoine set to work on the mountain in California where he was granted access to the largest telescope in the world. However, to avoid his taking up any coveted night-time viewing from seasoned astronomers, he was allocated a brief window around dusk. Doubtless, the director of the observatory, Horace Babcock, had sensed a whiff of genius in this young man! Over a few nights, the agitated images of nine different stars were recorded using an ultrafast camera. The measurements were analysed and published in 1972 (see Fig. 3.2). The images looking like bunch of grapes contained up to a thousand 'grains', in good agreement with the calculated size of the Airy disk associated with this large mirror. Antoine then published the apparent diameters of the disks of these nine stars, i.e., the angles they subtend when viewed from the Earth, based on a measurement of astonishing simplicity! (Gezari 1972).

The more poetic image of grape seeds originally attributed in French gradually gave way to the word *tavelures*, describing the pattern appearing on the skin of a fruit when its skin begins to age. But the phenomenon is referred to as 'speckle' in English.

The first breach had been made in the seeing barrier, but no image of the surface of these stars had yet been acquired since only the diameters of their disks were actually measured. There was still a long way to go. Many very short exposures were needed, a hundredth of a second at the most, so that

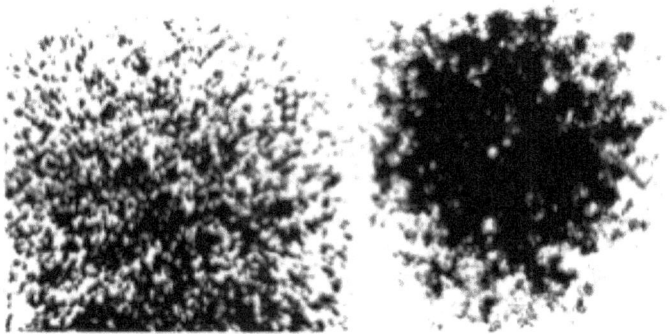

Fig. 3.2 Understanding the speckles in short snapshots of star images, photographed at the focus of the 200 inch Hale telescope on Mount Palomar. *Left*: The seeing disc on a 'point-like' star, with sharp speckles. *Right*: The image of the star Betelgeuse, showing much less contrasted speckles. Credit: A. Labeyrie

each exposure could make a snapshot of the atmospheric agitation. Then a certain amount of computation could extract some of the characteristics of the given star from this data. At the time, computing was developing fast, and it was particularly useful for these calculations. An ingenious colleague from Nuremberg and another of Antoine's unconditional admirers, Gerd Weigelt, came up with a subtle method, although it involved a lot of computation, which could produce true images reaching the diffraction-limited resolution of the given telescope. Thanks to this analysis of the speckle produced in both visible and infrared light, seeing was certainly no longer an absolute barrier, but the victory was not complete because such observations could only be made on bright objects. Better methods would have to be found.

Perhaps the reader will allow me a brief semantic interlude which will be useful in the following. It concerns the sense in which astronomers use the verb 'resolve'. In everyday parlance, it often has the meaning of finding a solution to a problem. But for photographers, opticians, and astronomers, the word has another meaning. For example, resolving a detail like a face on a blurred photo means reducing the blur sufficiently to make out the facial traits. The familiar notion of acuity or sharpness in an image should more correctly be called 'spatial resolution' or 'angular resolution' of the image. To improve the acuity, or rather the resolution of an image, means being able to better make out the details of the eyes and the grain of the skin, the thickness of a hair, and so on. Hence, to increase the resolution is to reduce the blur. Conversely, to reduce the resolution is just to blur so as to blot out the detail, as is done to protect image rights when people's faces appear in images posted online, for example. The level of resolution in any given image is to be sought in the characteristics of the instrument that produced it, or indeed in the path followed by the light before it reached the instrument.

To improve the resolution of images blurred by seeing, Antoine's method soon became the norm in the 1970s, and it produced all sorts of remarkable results!

Speckle in the Infrared

At the time, the promise of Antoine's first results was so clear to me that I could hardly ignore it. Not only did he break through the seeing barrier, but in 1975 he made another, even more fundamental breakthrough, using interferometry, so important that we shall devote a whole chapter to it. Blurring was not entirely unknown to me either. In 1966, as a young researcher studying the appearance of the outermost edge of the Sun's disk in the infrared, I had already been rudely confronted with blurred images at the Kitt Peak Observatory in Arizona, using the McMath–Pierce solar telescope. One astronomer particularly well informed about atmospheric effects and image sharpness was William Livingston, who patiently introduced me to these phenomena (Lynch and Livingston 2001). At the Californian Institute of Technology, he had studied under Horace Babcock, already mentioned above and soon to appear again. During the following decade, I would drift away from these first adventures in the world of blurred images to investigate infrared radiation from the Sun and then other stars, using telescopes carried aboard aircraft—NASA's four-engined Convair-990 in California and a twin-engined Caravelle in France, with the future astronaut Claude Nicollier. It was at this point that Daniel Rouan joined our research group in Meudon, freshly graduated from the *École normale supérieure*, and became one of the first to accompany us on our adventures in the infrared. We travelled this journey together for more than half a century, in the air,[2] in the basement of our research center, and at the top of the Andes, and he also joined me in our more recent attempts to improve the scientific education of young children in schools all over the world (a venture called 'Hands-on', and *La main à la pâte* in France). Daniel, whose sailing boat has been out so many times on the Mediterranean, has all the essential qualities of a sailor, in particular, reliability and precision.

---

[2]Daniel Rouan (2013) has told our astronomical and airborne adventures.

Then in 1976 I took a sabbatical year at the Kitt Peak Observatory. In the peaceful environment of its superb library, while my family, extended now to include four children, was taking in the beauty of the landscapes and sunsets of Arizona, I was busy trying to rethink Antoine's results and the prospects they might have in the infrared. Concerning the detection of this kind of light, I knew the techniques rather well, especially as the University of Arizona's Lunar and Planetary Laboratory, situated just a few hundred yards away, was one of the main centers of development in this field, thanks to the efforts of Frank Low and his student George Rieke. I thus built a very simple instrument which I called the *tavélographe infrarouge* (infrared speckle imager) and which was set up on the Mayall 3.8 m telescope at Kitt Peak. And so it was that we made the first speckle observations in the infrared!

When I got back to Europe in the summer of 1977, I took part in a summer school called *Infrared Astronomy* in the wonderful village of Erice, which looks down over the westernmost corner of Sicily, the three-cornered island known to the Greeks as Trinacria (Setti and Fazio 1978). The Ettore Majorana Foundation organises such meetings right through the summer, on every subject of current interest in physics and mathematics. Set up by Antonino Zichichi, a colourful Sicilian personality and physicist at the *Centre européen de recherches nucléaires* (CERN) in Genève, the center brings together both seasoned professors and emerging talent. During the following decades, the fifteen astronomers associated with our school on infrared in Erice would be leading figures from either side of the Atlantic, motivated by the common desire to explore the new world of the infrared in every way possible. Naturally, I gave a lecture on observation in the infrared, focusing on speckle. The youthful George Rieke came from Arizona to lecture on radiation from the galactic center, and presented the very first data. The long term follow-up to George's first results will be described in Chap. 7, where they will lead us to a black hole!

Once back in Meudon, my next problem was to decide which telescope we should use to pursue observations with our infrared speckle imager. Using the 3.8 m telescope in Kitt Peak and breaking through the seeing barrier, our imager had obtained the first diameter measurements (Chelli and Léna 1979) of a cloud containing a forming star, the so-called Becklin–Neugebauer object, in the constellation of Orion.[3] But access to this telescope was difficult because competition was stiff, and I was poorly qualified in the study of stars and galaxies, with little of relevance to attract interest. Regarding our astronomical

---

[3]https://en.wikipedia.org/wiki/Becklin--Neugebauer_Object.

missions aboard the Caravelle, although they were productive, they were by then considered too costly by the French research organisation, the CNRS, and we had to drop them.[4] I would have to look elsewhere. It was at this point that I decided to turn my attention back to the elimination of blurring.

In 1978, Jean-Marie Mariotti had just left the *École supérieure d'optique* in Paris, a breeding ground for the successors of Fresnel, Fizeau, and the likes of Labeyrie. He was a keen astronomer, dazzled by his encounter with Antoine, and spent many nights in the little dome that sits over the Sorbonne, in rue Saint-Jacques, Paris, sharing his enthusiasm with amateur astronomers. He was always very attentive in my lectures and I was impressed by his keen judgment and the rigour of his thinking, his capacity for work, hidden behind an apparent nonchalance, and his smiling sense of humour. As a research project, I suggested that he works on speckle, joining the little group I had set up with François Sibille at the Lyon Observatory and two young doctoral students, Christian Perrier and Alain Chelli. They all supported me admirably. Jean-Marie carried out three excellent observation missions to Arizona, then another in the Soviet Union where he used the 6 m telescope that had just recently been inaugurated at Zelenchuk in Crimea. He broke through the seeing barrier in observations of stellar envelopes, improving the resolution of these images by a factor of four.

At the time, my colleagues in Meudon had serious doubts about the future of these new techniques and I was unable to convince them to take Jean-Marie on in our research group, even though it specialised in the analysis of infrared data. But no matter, as Jean-Marie shared his first years of research between the University of Turin, where he could also discover his Italian roots, and the Lyon Observatory in Saint-Genis-Laval. The director of the latter was my friend Guy Monnet, a future key member of the ESO staff when buildng the VLT. He invited a young astronomer I had met at the Kitt Peak Observatory and who had not been accepted as a long term visitor by Meudon, in the turbulence of the years following 1968.[5] Steve Ridgway was an undisputed expert on a phenomenon which, in certain very specific cases, had already gone beyond the seeing limit, namely, the measurement of star diameters by lunar occultation.

---

[4] I shall not enter into the details of a European attempt to equip an Airbus A300 with a large telescope to match a similar project by NASA, thereby providing a stratospheric observatory with access to far-infrared radiation from space. This never got off the ground.

[5] This rejection by the Paris Observatory was easily forgiven 20 years later when the observatory solemnly awarded Steve Ridgway an honorary doctorate and appointed him as one of the foreign members of its visiting committee.

It is well known that the Moon has no atmosphere. Viewed from the Earth, it appears as a disk with a clear-cut edge or limb. When we look at this edge with an Earth-based telescope, it is no longer smooth, but slightly jagged due to the lunar mountains and it is no longer quite as crisp against the black background of the star-studded sky. This is the seeing effect. The Moon is in constant motion relative to the stars, because it is orbiting around the Earth. This means that, from time to time a star will disappear rather suddenly to the observer on the Earth because it has been occulted by the lunar disk. However, this disappearance is not strictly instantaneous because, as seen from the Earth, the star is not a dimensionless geometrical point, with no extension in space, but an extremely tiny disk! At the focal point of a telescope pointing at this star, just before the Moon occults it and then throughout the brief occultation, the astronomer observes a drop in brightness of the star that is not instantaneous. The time required for this, a tiny fraction of a second, is a measure of the star's diameter. Clearly, it is not affected by the seeing, because the phenomenon occurs outside the Earth's atmosphere—the light is propagating in vacuum when it encounters the lunar limb. This method had long been known to astronomers. It required a certain skill and some sensitivity. Steve had become one of the experts. His measurements in the near infrared gave star diameters of a few thousandths of a second of arc, some hundred times better than the seeing limit. With this background, Steve was ready to join our war on blur. A skilled experimenter, of few words and with sound judgement, his understanding of the subtleties of interferometry quickly became quite exceptional. With his wife Alisa, an inspired writer, he soon fell in love with France, its inhabitants, its landscapes, and its young researchers.[6] We have always met up with both of them, every year since then.

I found the period of unrest and discord following 1968 hard going, even exhausting, despite the many friendly relationships and interesting scientific activities that had sprung up during the decade. It was at this point that, quite unexpectedly, I was asked to collaborate with the European Southern Observatory (ESO), although at the time I knew almost nothing about it. This new collaboration marked an important and happy turning point for me. And a major turning point for the present story. I will thus return to it at some length a few pages on.

---

[6]After the attack which destroyed the World Trade Center in New York on 11 September 2003 and French opposition to the war in Iraq, the historian Philippe Joutard and his wife Françoise wanted to show that some Americans still loved France. In their book *De la francophilie en Amérique: ces Américains qui aiment la France* (Actes Sud, 2006), there is a contribution by Steve and Alisa Ridgway.

It was thanks to the collaboration with the ESO that I discovered the new European telescope inaugurated in 1977, a 3.6 m telescope in the Chilean Andes at La Silla, just north of Santiago. And my first thought was: why not use it for our work on infrared speckle and stellar envelopes? This looked like a good idea because the cost of our trips from Europe could be covered if our observation proposals were adopted by the committee in charge of allocating telescope time. Christian Perrier chose to do his national service as part of an international cooperation with an appointment in Chile, which meant that I could easily meet up with him. So there he was in La Silla, all ready to develop our project, which would then run for many years. However, despite the good results we obtained, it was clear that the acuity obtained by analysing the speckle would never be enough to unlock the secrets of star formation, and we were not the only ones in the world of astronomy to hold this opinion. Somehow, we would have to do better.

I enjoyed those long hauls aboard the Boeing 707, with all the stopovers, including those frequented by Saint-Exupéry when he worked for the French airmail company *Aéropostale*: Dakar, Sao Paulo, and Buenos Aires, then crossing the Andes within sight of Aconcagua, which stands at 6962 m, before diving back down to land on the plain at Santiago. I also enjoyed working with Christian, a graduate from the *École normale* and a very hard worker, who much later came a hair's breadth away from being selected by France as an astronaut. I subsequently had the same pleasure of working with Alain Chelli, another graduate from the *École normale*. I much admired his demanding stance on social justice, born of his family origins. In Chile, we French astronomers were always well received by an ebullient ESO engineer from Alsace, head of operations in La Silla (see Fig. 3.3). Always supported by his wife Sonia, Daniel Hofstadt, poet and environmentalist, became a good friend and would later be an able and respected negotiator between Chile and the ESO.

In a nutshell, the 1970s can be summed up by saying that, because a handful of astronomers like ourselves had grasped the importance of Antoine's work, we learnt to better understand the moods and agitations of the Earth's atmosphere and hence to get round some of the problems it raised. We were reasonably satisfied with this first step but we wanted to get true images more easily, without the blurring caused by the seeing effect, and at the maximum resolution allowed by the diameter of the telescope, just as an ordinary camera would. On the basis of everything that had been learnt about atmospheric effects throughout the decade, would we find a way to make that possible?

Fig. 3.3 La Silla Observatory, first ESO observing site in Chile, at 2400 m elevation, north of Santiago. *Left to right*: The 3.6-m telescope (1979) and the 3.58-m New Technology Telescope (NTT, 1982). Credit: ESO/José Francisco Salgado

Star Wars

In 1953, the 41 year old Californian astronomer Horace Babcock—the very same who, as director at Mount Palomar, would invite Antoine some 20 years on—published a paper with the prophetic title: *The possibility of compensating atmospheric seeing* (Babcock 1953). He noted that astronomers had long complained of this and that the problem "has generally been considered insoluble". Since Galileo, he could have added. He then wisely but firmly suggested "a method which, although subject to severe limitations, seems to offer in principle a means of compensating or correcting for the effects of atmospheric turbulence". A little later, in 1957, while unaware of Babcock's suggestion, the Soviet optical engineer Vladimir Pavlovich Linnik put forward a similar idea. But despite their promise of revolution, neither took the idea any further over the following two decades.

It was not only astronomers who were interested in correcting the effects on light of crossing the Earth atmosphere. At the time, the Americans and Soviets were engaged in the Cold War, a mad race toward mutual destruction. In their wargames, the military used a sinister unit to measure the number of potential casualties: the megadeath, or millions of dead. The principle of the laser had been discovered 20 years earlier and given them the idea of a 'death ray', one that they hoped would not remain forever confined to the realm of science fiction. Focusing the energy of a laser on one's enemy, vaporising his tank or his plane, was a bit like a modern version of what Archimedes was said to have done when he set fire to boats in the Roman fleet during the Second Punic War by focusing light on them with a huge mirror.

Ratified in 1967 just before the Apollo 11 mission won the race to the Moon that had been launched by President Kennedy, the Outer Space Treaty banned the use of spaceborne weapons of mass destruction but not satellites able to spy on potential enemies. To identify and monitor these satellites, what better than a powerful telescope based on the Earth's surface that could form a detailed image? So whether the problem was to focus a laser beam perturbed by atmospheric turbulence or sharpen up the image of a satellite, blurring was always at the heart of the matter and Horace Babcock's miraculous method for correcting blurring appeared once again on the programme.

Naturally, in secret, the US army, and no doubt also the Soviet army, copiously funded the research effort to reduce blurring. However, even in these periods of great tension, scientists tend to benefit from a certain freedom, and indeed this is essential for the exchanges involved in the scientific process. Indeed, at the beginning, the theoretical work analysing the effects of the Earth atmosphere on light was largely carried out by Soviet scientists, but it had been made accessible in the West. The same could not be said for the construction of instruments capable of correcting for the seeing effect. Astronomers, mainly in the United States, only heard vague rumours of this before, then throughout the period of office of Ronald Reagan, President of the United States from 1981 to 1989. Recall that, in 1981, Reagan launched what became known as his star wars programme. The aim was to develop equipment that could bring the Soviet Union to its knees in the race to exploit ever more sophisticated technology. And so the poor stars became a pretext for the Cold War. The war on blurring had been secretly confiscated by the military, their lasers, and their telescopes!

Although France had no longer been a member of the North Atlantic Treaty Organisation (NATO) since de Gaulle decided to leave in 1966, it remained one of the United States' main allies. This meant that some information nevertheless filtered through. The potential of Babcock's and Linnik's ideas had not escaped the notice of several French physicists, talented optical engineers paid by the French Ministry of Defence to look out for anything that might be useful to the military. One of these was Jean-Claude Fontanella, as astute as he was discreet, who would soon make a significant contribution to satisfying the aspirations of the astronomical community.

Europe and the True Stars

Let us go back for a moment to the rebirth of European astronomy. At the beginning of 1954, several German, British, French, Dutch, and Swedish astronomers met in Leiden in the Netherlands and came to the conclusion that astronomers in their respective countries should have access to a large optical telescope, something corresponding to the Hale telescope on Mount Palomar in the Northern Hemisphere. Combing the sky since its first light in 1949, this powerful instrument had discovered many new objects, notably extragalactic. A European instrument could be set up in the Southern Hemisphere, which had received much less attention than the other hemisphere up to then, although it contains for example the Large and Small Magellanic Clouds, two of the closest galaxies to our own. The sky in which the conquistadors "[…] Watched new stars rising, From the ocean floor into an unknown sky." (de Heredia 1981). After almost 20 years of political and financial deliberations and an aborted attempt to set something up in a South Africa torn apart by the ravages of apartheid, an international organisation was finally created in Paris on 5 October 1962 on the basis of a treaty between five countries. In the meantime, Belgium had joined the original group while the United Kingdom had preferred to remain in control of its own future projects.

The organisation of the European Southern Observatory (ESO) was copied from the one at the *Centre européen de recherche nucléaire* (CERN), set up in 1954. It was thus that the personnel of the ESO took up residence on the CERN site located near Geneva, on the border between France and Switzerland. Helped by scientists from across the Atlantic, the idea was to build a European knowledge center, so that researchers would have the tools to pursue their scientific ambitions, share ideas and resources, and finally recover from the aftermath of the two world wars. The ESO organised two important conferences in Geneva to lay down a framework for this common future. Since the founding programme of the ESO was to start building a 3.6 m optical telescope, the first conference, in 1972, concentrated on the design and the instruments that should equip it, such as cameras and spectographs. In France, the United Kingdom, and Germany, those countries still recovering from the destruction of the Second World War, where scientific research was just getting up and going again, other telescopes of similar aperture were also in the pipeline and their associated instrumentation naturally raised the same questions. At the ESO, the instruments were made, the main, 3.6 m mirror was polished, and the telescope was built at the selected site in La Silla, Chile. This telescope gathered its first light in 1977. It was the biggest that any European country had ever built, but it served only to catch up with the United States, after lagging behind for three-quarters of a century.

Having achieved this, the ESO, first with six member countries and soon with eight, with the addition of Italy and Switzerland, launched a second conference in the same year, 1977, to consider the next step. A new Director General, the Dutch and resolutely European Lodewijk Woltjer, had just been appointed (see Fig. 3.4). So what were the big questions in astronomy that a future instrument would have to deal with? An ambitious telescope, combining the resources of all the member countries, capable of bringing Europe back to the position it had occupied in previous centuries, capable of providing its youth, hungry for science and keen to develop their talents, with an alternative to crossing the Atlantic. The 1977 conference brought together everyone interested in this prospect. All the greatest astronomers of the day were there, from Europe, but also from the United States. The resulting project, a giant European telescope, was still only loosely defined and would take ten more years to materialise. This will be one of the main themes of the following chapters.

As it turned out, I was lucky. Just after the second Geneva conference, which I had not attended, I was surprised to find that I had been asked to join the Scientific and Technical Committee of the ESO which would oversee its day to day operations, and then later I would chair this same committee. And so it was that, at the beginning of the 1980s, I inherited a marvellous observation post, well placed to act in the battles to come!

Fig. 3.4 Lodewijk Woltjer, Director General (1975–1987), European Southern Observatory. Photo by Pierre Léna

The Birth of Adaptive Optics

Horace Babcock, the visionary of 1953, came from California to attend the 1977 conference at which the outlines of the future European telescope would be spelt out. Would he defend his idea for correcting the blurring of images, whose authorship no one would contest? Did he really believe in it himself? We may well wonder since the only time he spoke on this subject during the conference was to emphasise, and quite rightly, the importance of the choice of mountain, to reduce the seeing effect as far as possible by paying attention to the quality of the atmosphere above it. He did not mention the possibility of correcting for the detrimental effects of atmospheric turbulence. In fact, it was a French physicist, Gilbert Bourdet, involved in work with the French Ministry of Defence and not part of the astronomical community, who noted rather discreetly that Babcock's idea could be put to use.[7]

A year after this conference in Geneva, I took up my position as chair of the Scientific and Technical Committee of the ESO. So there I was, working on this great project for a new telescope. Given what we knew about speckle and given Antoine's first results using interferometry, as will be described in Chap. 4, the possibilities for countering image blurring seemed sufficient to persuade the Director General to call a further conference in 1981 to discuss what had by then become known as the problem of high angular resolution (Ulrich and Kjär 1981). This generic but somewhat mysterious term specifies a clear objective. At the wavelengths of visible or infrared light, we wanted to explore all the possible ways of getting round two barriers preventing us from obtaining perfectly sharp images. First, the seeing barrier, and then, assuming that there was a way to get around this first obstacle, the diffraction barrier. Indeed, diffraction of light by the edges of the large telescope mirror is what causes the residual blurring introduced by the presence of the Airy disk. Our dream was to replace the seeing limit of one second of arc by a much smaller one, perhaps ten, a hundred, or even a thousand times smaller, thus improving the sharpness of our astronomical observations by the same factor!

From then on, the way to go beyond the first obstacle would be called adaptive optics, while our method for getting round the second would be called optical interferometry. Actually achieving these methods, step by step, would take 20 years, and it will occupy much of the following story! Throughout this process, the two approaches would share many common features and each

---

[7]"One possible, highly effective way to improve the quality of instruments forming an image in the presence of atmospheric turbulence is real-time control of the phase [of the wave]": Bourdet (1978).

would often throw light on the other. Success with the first would be essential if we were to triumph over the second. So it is only for the purposes of simplicity that I have separated the two issues. In the present chapter, I shall thus pursue the first, adaptive optics, over the first of those two decades, and then in the next chapter, I shall begin to discuss the second, optical interferometry, over the same period.

At the ESO, I met the young German engineer Fritz Merkle, an excellent optician. He had just been taken on after finishing his doctorate. We got on well from the start. He was well organised, methodical, friendly, and full of energy, and he was afraid of nothing. We would work together for the next 10 years, him in Garching near Munich where the ESO had just taken up residence, and myself usually in Paris. Our aim was to show that Babcock's idea—compensating atmospheric seeing, to use his own words—could really work, and we were ready to put our money where our mouths were. In France, intrigued by Gilbert Bourdet's remarks at the 1977 conference, I discovered that the *Office national d'études et de recherches aérospatiales* (ONERA), under the aegis of the French defence ministry, was interested in using adaptive optics, probably to build a tactical weapon that involved the concentration of laser beams.[8] I was already acquainted with ONERA, having worked with them 20 years earlier, supported by the excellent physicist André Girard, when I began my first work on infrared astronomy. There, Jean-Claude Fontanella, another graduate from the *École supérieure d'optique*, was also interested in our discussions, and indeed, he fully understood what was at stake. A research laboratory, the *Compagnie industrielle des lasers* (CILAS),[9] was working with him on the subject in Marcoussis near Paris. There I met the leader of the research group, Michel Gaillard, another visionary, and an excellent physicist and astute politician, open-minded enough to subscribe immediately to our project. A close-knit team was soon set up between this industrialist, Fritz at the ESO, Jean-Claude at ONERA, and myself at my research center in Meudon. Shortly, afterwards, CILAS supplied us with our key component, a flexible and deformable mirror. In the meantime, I explained to the military that the performance requirements for astronomical applications would be much more demanding, and hence much more interesting than those of the army, but that I could not give up our freedom to publish. The French defence ministry took the point gracefully.

---

[8]ONERA is the French national aerospace research centre. A good presentation of the role played by this organisation in developing adaptive optics for applications in astronomy can be found at the web site www.onera.fr/actualites/loptique-adaptative-a-lonera-de-ComeOn-a-le-elt.

[9]CILAS-Alcatel, https://fr.wikipedia.org/wiki/Compagnie_industrielle_des_lasers.

The time has come to explain in some detail the basic idea put forward by Babcock, the principle of an adaptive optical instrument. Light from a star reaches the telescope in the form of a wave. As Newton had understood, the image is made up of "those many trembling Points confusedly and insensibly mixed with one another". In other words, instead of being planar, the wavefront has been battered into an irregular shape as it crossed the atmosphere. It is these fluctuating lumps and bumps that cause the multitude of trembling points in the image. But as discovered by Antoine at Mount Palomar, the instantaneous image retains a memory of—more precisely, information about—what was contained in the original wave. The presence of this information is revealed by the rapidly shifting, disordered fine detail which we called speckle.[10] So the image can be analysed and, thanks to this residual information, we can work out exactly how the wave was 'knocked around' as it flew through the atmosphere. This analysis must be carried out very quickly since the state of the atmosphere can change on time scales of a fraction of a second.

So what exactly happens to the light wave as it crosses the Earth's atmosphere? After all, when it first reaches the atmosphere, it will be a perfectly plane wave, while it arrives at the telescope in this battered state. Let us take a moment to consider the surprising phenomenon of atmospheric turbulence. Lean over the parapet of a bridge that crosses a fast-flowing river and look at the water downstream of the pillars of the bridge. What we see are quite large whirlpools getting swept away by the current. These whirlpools soon break up into smaller ones which themselves then break up, and so on and so forth, until the friction between them makes them disappear. In the presence of an obstacle like the pillar supporting the bridge, the beautifully ordered flow of the water molecules upstream is transformed into a chaotic agitation of the fluid. This is what we call turbulence, in this case hydrodynamic. This chaotic phenomenon is hard to describe mathematically. It was the great Russian mathematician Andrey Kolmogorov who, in 1941, formulated the first accurate description of this hierarchy or cascade of turbulent whirlpools.[11]

---

[10]Astronomical photos of stars with exposure times much longer than a tenth of a second mix up the speckles into a blurred spot in which no more detailed information can be discerned. But some observers, like the astronomer Paul Couteau in Nice, had a reputation for making sketches of much better quality than these photos, even making out a double star where the photos show only a single splotch of light. This is due to their ability to catch hold of the snippet of information which is retained in the speckle, but disappears in the long-exposure image. This was indeed how Jean Rösch had been able to describe a fleeting bunch of grapes structure.

[11]In 1922, L.F. Richardson adapted a famous verse of the satirical poem *On Poetry: A Rhapsody* by Jonathan Swift, author of *Gulliver's Travels*, giving this elegant description of turbulence: "Big whirls have little

Fig. 3.5 On the VLT platform, looking to the east, three of the egg-cup shaped telescopes are illuminated by the setting Sun. Clouds are indeed very rare above the site in Paranal, but here strong winds from the Andes have created steady atmospheric turbulent eddies, made visible in the cloud shapes. Credit: ESO/Pierre Bourget

The same phenomenon occurs in the air when the wind flow encounters an obstacle like a mountain (see Fig. 3.5). The resulting whirlwinds, which can last 10 or 20 min, mix volumes of air at different temperatures and on smaller and smaller length scales. As the air is transparent, this disorder is invisible to the eye. It can sometimes be revealed by an aircraft contrail which begins to 'fluff up'. But light itself moves faster in hot air than in cold, so it is very definitely affected by such a disordered medium. In order to reach the various points on the telescope mirror, the light rays follow different paths which will have required different travel times. From one point on the mirror to another, the wave may have arrived either slightly before or slightly after. The wave front, initially plane, thus arrives warped and buckled.

An adaptive optical instrument brings together three features. 'Straightening out' the wave works rather as a boilermaker would rectify the shape of a sheet of metal, delaying the wave over any part of the mirror where it arrives

---

whirls, That feed on their velocity, And little whirls have lesser whirls, And so on to viscosity." (Richardson 1922).

ahead of schedule, and advancing it wherever it arrives late. The first feature is therefore to analyse the wave front, i.e., to analyse the lumps and bumps on it and measure them. The longer the wavelength of the light, the fewer the regions of the wave front that will need correcting, a point that favours working with infrared light. The second feature is a deformable mirror, that is, a small flexible mirror. This deforms very quickly to make the necessary correction, then reflects the wave in such a way as to exactly counter the unwanted atmospheric effect. Once again, the shorter the wavelength, the faster the correction must be made. And finally, the third feature, a computer program which calculates, still in real time, exactly how the deformable mirror must deform in response to what the analyser has measured. The whole thing thus makes use of a feedback loop. In real time, the analyser measures, the computer calculates the correction, the flexible mirror applies the correction to the wave which reflects from it, and the loop is complete.

There is nothing particularly difficult about this once the idea has been formulated. The only thing is to have a computer that can work things out that quickly, and this is precisely what Babcock didn't have! In principle, this system can work at any wavelength of the distorted wave. Two properties explain why we chose to work with infrared light, as just mentioned: fewer mirror elements to be deformed and more time to make the corrections. Today, 30 years on, adaptive optics is used the world over and has become considerably more sophisticated, but it still features the three basic components just presented— a wave front analyser, a deformable mirror, and a computer to control the feedback loop—and it is more often applied to wavelengths in the near infrared (see Fig. 3.6).

Lo Woltjer, the Director General of the ESO, was a man of great culture. An enthusiastic astronomer, often skeptical about human beings, he had deep convictions, believed firmly in the tradition of European excellence, and asked thought-provoking questions, but always listened to engineers and knew whom he could trust. He was also a leader who knew how to run the show, sense what was in the air, and make ambitious but reasonable decisions, as far as these two terms were not downright contradictory. He was determined that the VLT should succeed. It was his life's great work. He gave us his full support, and by 1985, a prototype adaptive optical system called ComeOn was under joint study by the ESO, the Paris Observatory, and ONERA. This was fortunate because, with the exception of Michel Combes, who would later become the director of the observatory, my French colleagues at the observatory had little

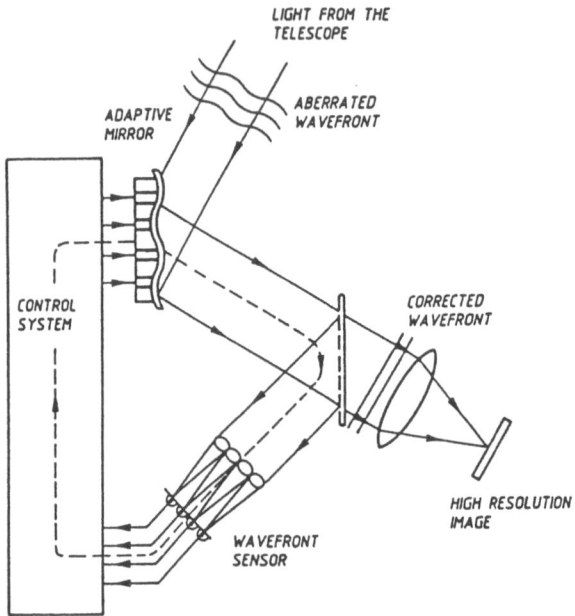

LIGHT FROM THE
TELESCOPE

ADAPTIVE
MIRROR

ABERRATED
WAVEFRONT

CONTROL
SYSTEM

CORRECTED
WAVEFRONT

HIGH RESOLUTION
IMAGE

WAVEFRONT
SENSOR

Fig. 3.6 This schematic drawing illustrates the three main components of any adaptive optics system for astronomical use: the deformable mirror, adapting its surface to the wave front deformation, the wave front sensor, measuring these deformations, and the control system, closing the feedback loop from the sensor to the mirror. Sketch by Pierre Léna

faith in our plans to remove the blur.[12] For them, the whole thing seemed crazy, a wild dream that was costing already scarce resources.

They were not the only ones to be skeptical. As mentioned above, in the United States, astronomers were focused on the prospects offered by the Hubble telescope which, being spaceborne, would already escape the blurring caused by the Earth's atmosphere, and the only thing that came their way were the occasional snippets of information about the efforts of the military to build an adaptive optical system, largely held secret. However, we remained in close contact and I often returned from the US convinced that, with the scant resources at our disposal, we in Europe were certain to lose first place in

---

[12]The activities of the *Laboratoire d'études spatiales et d'instrumentation en astrophysique* (LESIA) at the Paris Observatory in Meudon are described clearly and at length at the web site www.lesia.obspm.fr/ Historique-de-l-optique-adaptative.html. Fortunately, this account shows no sign of the initially rather skeptical attitude as I remember it personally.

the race, because some of my colleagues in the US had clearly grasped what was at stake and they had all the financial support they needed.

These transatlantic exchanges sometimes gave rise to astonishing comments. Although an excellent physicist, an astronomer friend of mine to whom I had just explained the principles of the instrument we were designing, made the remark: "It's too good to be true!" Hence the title of the present chapter. And it genuinely is good, almost too good, and yet it works! Another, more chauvinistic colleague, seriously made the following rather out-of-place remark: "Pierre, if you were right, we should have found it first!" No comment!

One Night in Provence

To build and test the prototype ComeOn, Fritz had apparently unlimited energy. However, we needed keen and creative young colleagues to help us, and fortunately, my position as professor at Paris Diderot University—then called the University of Paris VII, in the not very suggestive nomenclature inherited from the Edgar Faure law of 1968—brought me into contact with many enthusiastic candidates who had opted for a doctorate in astronomy. Our challenge could tempt the more daring among them, who, at the age of twenty-two, were ready to take a few risks. They included Pierre Kern, brought up in the tradition of French optics, and also François Rigaut (see Fig. 3.7). ONERA was fully committed to the project as overall designer, with Jean-Claude Fontanella, who had for his part recruited the young Gérard Rousset, a rigorous, methodical, and discreet engineering graduate from the *École centrale*. It was the latter who designed the wave front analyser, which played as critical a role as the deformable mirror. Several other engineers played key roles: Pierre Gigan at the Paris Observatory, Corinne Boyer and Pascal Jagourel at ONERA, and Jean-Paul Gaffard at CILAS all contributed to designing and building the instrument. Our little ComeOn could sit on a modestly-sized optical table. We wanted to carry out the first tests in France and chose to set up ComeOn at the *Observatoire de Haute-Provence* (OHP), pride and joy of the CNRS, located in the fragrant landscapes of the Lubéron, near the village of Saint-Michel l'Observatoire. We chose this as a matter of convenience because the focal area of the 1.52 m telescope was conveniently fixed with respect to the ground.

The date of the test was fixed for the first few days of October 1989, and in mid-September, I went down with the team to set things up, awaiting the first series of observations. But I was suddenly caught up in an impossible dilemma. For in 1989, I was elected president of the *Société française de*

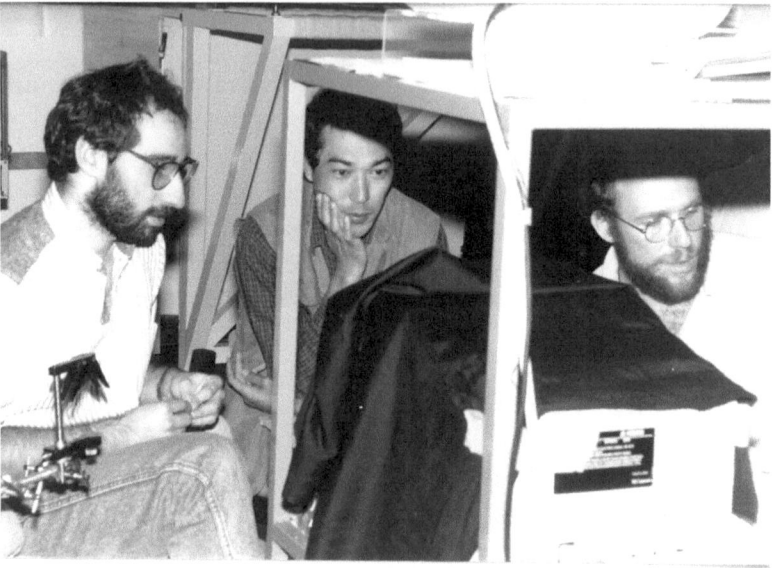

Fig. 3.7 Provence 1989. Three young astronomers, busy installing the ComeOn instrument at the *Observatoire de Haute-Provence*. *Right to Left*: Pierre Kern, Almas Chalabaev, François Rigaut. Credit: Observatoire de Paris/LESIA

physique. Now, nobody would understand why the president was not present at the *École normale supérieure* in Lyon on 25 September to welcome Andrei Dmitrievich Sakharov and his wife Elena Bonner, the society's special guests (see Fig. 3.8). My meeting with this man was an unforgettable experience. He was the theoretical physicist who had designed the Soviet's hydrogen bomb under Stalin, but later, paying the price with his own freedom, he had become a pacifist and staunch defender of the rights of the oppressed, and in 1975, received the Nobel Peace Prize. Throughout that week, alongside the mayor of Lyon, Michel Noir, and the former defence minister, Charles Hernu, deputy mayor of Villeurbanne, we attended seminars, meetings, and banquets in honour of our guest, who then returned to Moscow.[13] Andrei Sakharov didn't survive the violence of the conflict that opposed him to Mikhail Gorbachev at the Duma, where he had been elected to parliament. In a televised scene, to the stupefaction of the master of the Kremlin, he requested the abolition of Article 6 of the Soviet constitution, the one establishing single party rule. He

---

[13]An account of these events can be found in: Andrei Sakharov, *Science et Liberté*, Editions de Physique, Paris, 1990. Further references can be found in Yves Quéré, *Un coquillage au creux de l'oreille*, Odile Jacob, Paris, 2018.

Fig. 3.8 Science and Freedom in 1989. While the blur was being defeated at the *Observatoire de Haute-Provence*, Andrei Sakharov was being made Doctor Honoris Causa of the Université Claude-Bernard, not far away in Lyon, by its president Jean Zech. Photo by Pierre Léna

died on 14 December 1989, 34 days after the fall of the Berlin Wall, and just 3 months after his visit to France. Doubtless this has little to do with the rest of my story, but it is important, as I see it, to pay homage to this man.

After the experience of this historic visit, I had to return to the university and welcome the new intake of astrophysics students, get ready for the next meeting of the ESO Council of which I was now a member, and also prepare the annual exhibition of the *Société française de physique*. Already too much for one person, and I am sad to say that I was unable to attend the first successful operation of ComeOn, on the night of 12 to 13 October 1989. As a result of this observation, the whole team signed the photograph attesting to our shared achievement. On the image given by the infrared camera, a double star called γ_2 Andromedae was perfectly resolved.

The star γ in the constellation of Andromeda is actually a multiple star system in which four stars orbit around one another. The companions γ_1 and γ_2 are separated by 10 s of arc, which is ten times the seeing disk, whence this bright pair has been known to astronomers since 1777. However, the companion γ_2 is itself a double star and its two components $\gamma_2 A$ and $\gamma_2 B$ are very close, in fact separated by only half a second of arc at the time of our observation. The seeing conditions thus mean that they are impossible

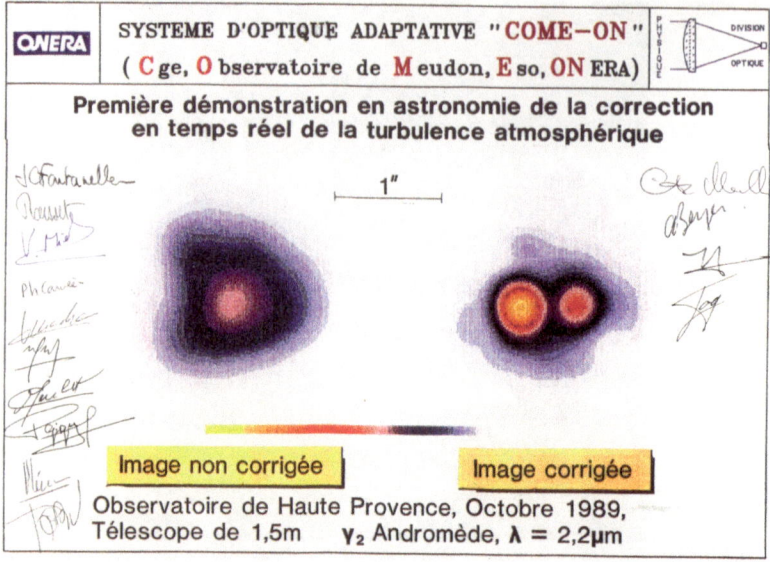

Fig. 3.9 An historical document. At the end of the night, astronomers and engineers celebrated the success of the ComeOn instrument, by signing this picture. *On the left*, the blurred image of the double star γ_2 Andromedae. *On the right*, the adaptive system is active. The two companions are clearly separated. Image provided by Pierre Léna

to make out within the fuzzy disk. Indeed, the binary nature of the star γ_2 was only much later discovered using a spectroscopic method,[14] rather than by direct imaging. And the star γ_2A is yet another double star, an even closer binary system, also discovered by spectroscopy in 1959. The separation there was so small that we were still unable to reveal it with ComeOn.

Pointing our telescope at γ_2 and applying our adaptive optics system, the fuzzy disk was instantaneously transformed into two beautiful distinct images (see Fig. 3.9), each perfectly circular: the star γ_2 had been split into two! What's more, the image of each companion was surrounded by its Airy disk, showing that, once we had got past the problem of atmospheric blurring, all that remained was the diffraction due to the telescope itself, this determining the diameter of each image. In my notebook for that day, following the phone call informing me of the result, I simply wrote: "Adaptive optics at the OHP. A great success."

---

[14]The motion of the companions A and B, in orbit with a period of 62 years about their common center of mass, produces an alternating shift of the spectral lines produced by these stars, toward the blue for the one moving toward the Earth-based observer and toward the red for the one moving away. All the spectral lines are thus split.

On 26 October, under the title *Adaptive optics shown on French telescope*, the journal Nature described in detail how "a dancing, bobbing image was reduced to an almost stable spot", whose size came close to the theoretical limit imposed by diffraction, i.e., 250 milliseconds of arc at this infrared wavelength and for this modest telescope. The article continues: "the gamble taken by ESO in 1985 seems to have paid off". By then, the ESO had a new director who had informed the British journal without even consulting us, mentioning no one but himself and, fortunately, Fritz! The small-mindedness of some important men. This detail was promptly rectified in the ESO's journal Messenger (Merkle et al. 1989) and did nothing to spoil our pleasure. Our own publication soon followed with the modestly triumphant title: *First diffraction limited astronomical images obtained with adaptive optics* (Rousset et al. 1989). As is usual in these cases, the first to be named on the list of authors, Gérard Rousset, was chosen among the youngest in the team. Thirty years on, Gérard is now professor at Paris Diderot University where he directs the high angular resolution team in our laboratory in Meudon and is preparing the adaptive optical instruments for the future 39 m European telescope.

It is interesting to note that, in the United States, the news of our success spread round the observatories like wildfire, doubtless causing a certain bitterness in a country where it is considered natural to come first. And just 3 weeks after our success, on the 9 November, the Berlin wall finally came down. The Cold War was over and our astronomer colleagues now had sound arguments for requesting the declassification of what had until then always been a closely guarded military secret. Two years later, an article was published in the United States which disclosed some of the secret, non-astronomical work that had been carried out until then. Questions and solutions could now be openly discussed with our competitors across the Atlantic, as would normally be done in the world of research. The American military and industrial machine was in fact very slow to react to the end of the Cold War, since it was not until 2011 that the airborne laser (ABL) programme, a plan to install a giant 1 megawatt chemical laser equipped with adaptive optics aboard an aircraft, was finally abandoned after spending 5 billion dollars over 16 years[15]–something like ten times the cost of our own European VLT!

---

[15] I learnt this interesting piece of information from Olivier Lai at the *Observatoire de la Côte d'Azur*. We shall meet him again later in our story. He describes this in more detail in his habilitation thesis *Vingt ans d'optique adaptative sur Mauna Kea*, University of Nice, 2017.

Great Prospects for Adaptive Optics

Our instrument ComeOn was very soon set up in Chile at the focal point of the 3.6 m European telescope on the mountain of La Silla where, 10 years previously, I had spent many nights with my infrared speckle imager, in my early struggles to beat the blur. On this same mountain, we were keen to demonstrate the advantages of these new images (Rigaut et al. 1991), since we could now achieve the same angular resolution as the Hubble space telescope, which was flying round the world once every 90 min. We improved our instrument and obtained a profusion of results at a resolution of a hundred milliseconds of arc. Little by little we began to apply our imaging technique in every field, such as orbits of until then unresolved binary stars, the surfaces of asteroids and even of Titan, one of the natural satellites of Saturn, and regions of star formation studied by a shrewd young graduate from the *École normale*, Fabien Malbet, who had joined our group to work with the instrument ComeOn.

In 1990, Pierre Kern presented his doctoral thesis. It was entitled *Adaptive optics and large telescopes*, a world first! He had just become a father. From then on, he never left us, making his name at the University of Grenoble and setting up a small adaptive optics company, before taking up a position as technical director at the CNRS where his understanding of the technological challenges remains invaluable. On the basis of our first results, François Rigaut was next to defend a brilliant doctoral thesis which immediately earned him a well deserved reputation as an adaptive optics specialist—there were hardly any around at the time! He would soon receive many tempting offers from the United States, then Australia, and before long he was lost to France, although not of course to science. Alas! He would not be the only one. We seem sometimes to be unable in France to provide good working conditions to this young generation.

At the beginning of the 1990s, adaptive optics had proved its worth. The atmosphere and the seeing effect had been almost completely overcome. Images could now be 'deblurred'. A particularly apposite German neologism describes what adaptive optics can achieve: the verb *entfunkeln*, which literally means the opposite of 'twinkle'. Adaptive optics 'detwinkles' the stars which atmospheric turbulence has caused to twinkle.[16] In the United States, the idea was gaining ground and little by little earlier military and industrial research

---

[16] The Supernova Planetarium & Visitor Center in Garching, right by the ESO headquarters, gives the visitor an excellent presentation of contemporary astronomy and the ESO observatories. This verb *entfunkeln* is used there to present adaptive optics.

was being declassified, so in 1992 half a dozen observatories had begun to set up adaptive optics devices on their main telescopes (Beckers 1993).

By the end of the twentieth century, the largest optical telescopes, with primary mirrors between 3 and 6 m, would be superseded by a new generation with apertures of 8–10 m. Since the size of the Airy disk decreases in proportion to the diameter of the primary mirror, this new triumph over the blurring caused by diffraction that we had been longing for right through the previous decade would now become possible by equipping these giant telescopes with adaptive optics. After an improvement by a factor of five in the resolution due to our victory over the seeing effect, a further leap forward by a factor of two or three had now become possible with these new telescopes.

The European project which we shall describe in detail in Chap. 5 could now be built upon a firm foundation, allowing us to take this next step and reach a resolution of 50 thousandths of a second of arc in our infrared images, twenty times better than the seeing limit!

Nothing could stop us now. Having broken the seeing barrier by applying adaptive optics, the next target was the diffraction limit of the telescope itself. It was time to make way for the champions of interferometry!

4

The Quest for Sharp Images: Interferometry

Every great action is extreme when it is undertaken—Stendhal, *The Red and the Black*

How can we represent light? What is the nature of this apparently elusive entity? What exactly is this phenomenon produced in the Sun, carrying energy to provide warmth for us, and which can cross the empty space between galaxies and even reach us from the most distant moments of the Universe just after the Big Bang? What are these ranges of quantities and qualities used to describe it: energy, frequency, wavelength, and colour? Different descriptions of the same thing: the light we perceive with our eyes, from violet to red; the warmth of the infrared that heats our skin and the ultraviolet that can burn it; other forms of light that our senses cannot detect, such as very low energy radio waves and very high energy X-ray and gamma-ray photons. Why did it take so long to invent this special source of light we call the laser, which has become so commonplace over the last half century?

We will not be able to answer all these fascinating questions here. Instead, let us take a look at four particularly important luminaries who contributed to our present state of knowledge between 1802 and 1975. Having understood certain fundamental properties of light, although not all, they each made essential contributions to solving our present problem, the quest to obtain sharp images. What set them apart was the originality of their ideas, the perspicacity in the way they scrutinized the natural world around them, and the boldness of the experiments they were inspired to carry out.

© Springer Nature Switzerland AG 2020
P. Léna, *Astronomy's Quest for Sharp Images*, Astronomers' Universe,
https://doi.org/10.1007/978-3-030-55811-6_4

Four Breakthroughs Over the Centuries

Three of our explorers were the British polymath Thomas Young, the French physicist Hippolyte Fizeau, and the American physicist Albert Michelson. Thanks to them, by the beginning of the twentieth century, a solid foundation had been laid for understanding the nature of light. In what follows we shall therefore take a little time to talk about their work. Then, in 1975, we come to our fourth explorer, the French astronomer Antoine Labeyrie, who subsequently occupied a chair at the *Collège de France*. Over this same period, the 1970s and 1980s, once adaptive optics had come into being, a second line of attack was opened on the problem of blurred images. And this, too, would soon be rewarded by a second victory. So let us get on with our story. Then later, with the observation of exoplanets and a quite singular black hole, this tale will take us to Paranal.

A Young Man with Imagination

To the end of his life, the great scientist Isaac Newton had struggled to understand the natural phenomenon of light. He wanted to explain all its manifestations: the colours, the rainbow, refraction, and diffraction by a hair as observed by Francesco Maria Grimaldi in the seventeenth century and later by others. Naturally, he hoped to extend his conception of matter, which had been so successful in establishing his law of universal gravitation. He was thus convinced that light must be made up of tiny particles, moving in straight lines along the light rays. But with this conception of light, he was never able to explain the observations he himself described so admirably, such as the coloured rings that carry his name, produced when two illuminated glass surfaces of different curvature are brought into contact.

And so it was left to the youthful British polymath Thomas Young, physicist, linguist, doctor, and musician who, in 1802, carried out the experiment that made him famous, an experiment that every student repeats today as an essential step in their understanding of nature. Many in fact consider this experiment (and its intepretation) as one of the most beautiful physical demonstrations of all time.[1] Using a mirror, Young reflected a beam of sunlight onto a screen in which a hole had been made with a pin. The emerging thin ray of light was then sent toward a piece of opaque card in which two fine

---

[1]The double-slit experiment, Editorial, Physics World, Institute of Physics, 1 September 2002.

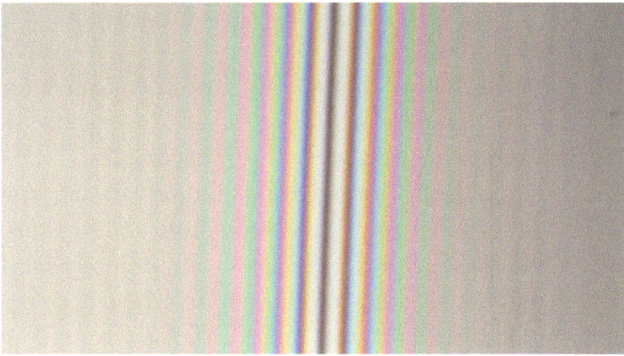

Fig. 4.1 Young's fringes, obtained from a white light source. The spacing of the fringes depends on the wavelength of the light, blue being shorter than red. Credit: P. Chavel, Institut d'Optique-Graduate School

slits had been cut, less than a millimeter apart ("one thirtieth of an inch"). On the wall behind, used as a screen, Young observed that fringes were produced, i.e., alternating stripes of light and darkness, whereas everyone would have expected a uniform illumination. Being a musician, Young compared this observation to the variations in the intensity of sound that can be heard when two sounds of identical or very similar frequency are produced together. He also gave a name to this phenomenon, calling it interference, using it for the first time to refer to the physical phenomenon we are concerned with here.[2] These interference fringes are so important in the present story that it will be useful for the reader to actually visualise them (see Fig. 4.1).

It is said, and it may or may not be true, that the explanation came to Young in the following manner. While studying at the famous university in Cambridge (UK), he was once staring at the calm surface of water in an ornamental pond in one of the many gardens. Then two swans came down, slipping across the surface and stirring it slightly. The wrinkles caused by their approach propagated across the surface and came to meet. Two wave peaks would add up to make a slightly bigger wave, whereas a peak meeting a trough would cancel one another, and two troughs would combine to produce an even deeper trough. The phenomenon we call interference results from these additions and subtractions of pairs of waves. In a second step, Young deduced something about the nature of light from the presence of these fringes.

---

[2] Mollon (2002). This article is followed by many interesting contributions, further discussing the impact of Young's experiment on the whole of physics over the next two centuries.

He concluded that light could not be made up of particles, as Newton had supposed, but had to be a wave.

Now, sound waves propagate through air, and in fact no sound wave can cross a vacuum. Waves on the surface of water also require this material medium, water, in order to propagate. But what medium did light waves require for their propagation? Young could not provide an answer to this question, and it would bother physicists throughout the nineteenth century. For want of a better idea they called it the aether, an infinitely subtle substance that was assumed, whatever it was, to fill the otherwise empty space between ourselves and the stars, since light does indeed reach us from the stars. A final answer to the mystery of the aether was of course found, but it is not essential to our purpose here.

Apart from this key result, Young was the first to measure the wavelength of light rays visible to the naked eye, that is, lying between violet and red, at the two ends of what we call the spectrum, as seen in a rainbow, for example. The wavelength can be deduced from the spacing of the observed interference fringes. Better still, Young went back to the measurements of another phenomenon, Newton's rings, that Newton had published without being able to provide any explanation. Young, however, could indeed explain them as being due to interference, whereupon he could immediately use measurements published by Newton to deduce the wavelengths of blue and red light. They are roughly 0.3 and 0.7 μm, hence less than a millionth of a meter. These values would lie at the heart of attempts to remove blur some half a century later.

The wave nature of light that Young had corroborated in this way, just before the Napoleonic wars, the Battle of Waterloo, and the Treaty of Vienna transformed the contours of Europe, went on to inspire physics throughout the nineteenth century.

"And Incidentally, We May Expect ..."

Armand Hippolyte Fizeau was the son of a doctor, himself an acquaintance of the great René Laennec, the doctor who, in 1816, invented the stethoscope to improve the sensitivity of auscultation. The latter's son-in-law, Adrien de Jussieu, was a botanist and son of Antoine-Laurent de Jussieu, the first director of the Natural History Museum, founded in Paris by the Convention in 1793.[3] Hippolyte was immersed in science from birth thanks to his family

---

[3] The National Convention (*Convention nationale*) was the first government of the French Revolution.

background. He soon took up physics and became a very capable experimenter. Indeed, he was brought up among the ideas and experimental demonstrations first elaborated by Young, then developed by two great French scientists, Augustin Fresnel and François Arago. Experimenting on light himself, in 1848, he discovered the effect that carries his own name, the Doppler–Fizeau effect. The other name here refers to the Austrian physicist Christian Doppler, who demonstrated and correctly interpreted the same phenomenon when produced with sound waves. When a source of sound or light is moving relative to an observer who measures the wavelength, the latter—and hence also the frequency—is modified, becoming a function of the relative velocity of the source and observer. Measurements show that the wavelength increases when the source is moving away from the observer, and decreases when the source approaches the observer. This is a fundamental result in astronomy, because it can be used to determine the speed of the stars relative to the Earth. Hippolyte Fizeau did many other things, for he was the first to make an accurate direct measurement of the speed of light, in Paris in 1849. He is one of a select group of scholars whose name features on the great cast iron frieze around the first level of the Eiffel tower.[4]

In 1851, already a leading expert on the properties of light, Fizeau had a quite revolutionary idea, inspired by Young's interference fringes. This idea is developed in detail in his handwritten notebooks, conserved in the archives of the French Academy of Sciences (see Fig. 4.2).[5] Going back to Young's experiment, he noted, as Young himself must have done, that with a very small light source, there is greater contrast in the resulting fringes. As the size of the source increases, the contrast is reduced and finally disappears. There is thus a connection between the angular size of the source and the contrast in the fringes.

The important step taken by Fizeau was to ask what would happen if the light source were a star and if the screen with the two holes—or slits, to transmit more light—were placed in front of the telescope mirror. These slits would thus isolate two pieces of the light wave and one ought to observe fringes at the focal point of the telescope. If the angle subtended at the Earth's surface by the diameter of the star, which we shall call the apparent diameter of the star, were to increase, then the fringes should fade and disappear. Hippolyte Fizeau kept this idea to himself, and it was only seventeen years later, in 1868, that with a certain modesty, he formulated it publicly. He did so when he was

---

[4]See the excellent biography written by J. Lequeux (2014).
[5]Fizeau's handwritten notebooks. *Académie des sciences, Institut de France.* Fonds 64J, Hippolyte Fizeau. Dossier 9.01.

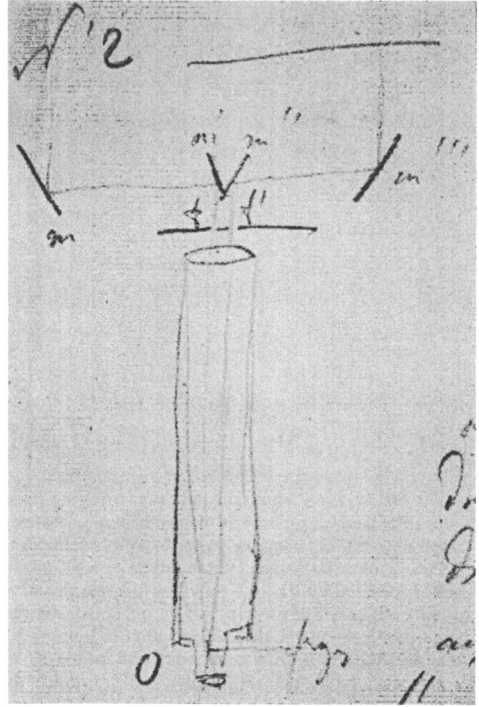

Fig. 4.2 In 1851, on a crude but historical sketch, fortunately kept in archives, Hippolyte Fizeau outlined his brilliant idea for measuring the dimension of a star with two apertures in front of a telescope. This scheme is exactly the one which would be used by Albert Michelson on the Hooker telescope at Mount Wilson in 1891. Credit: Archives of the *Académie des sciences*

awarded a prize in the great hall of the *Institut de France*, headquarters of the Academy of Sciences, under the watchful gaze of Corneille and La Fontaine. This text is so fundamental to our story here that it is worth citing in detail:

> For most interference phenomena, like Young's fringes [...], there is a remarkable and necessary relationship between the size of the fringes and the size of the light source. It is such that the fringes, faint as they are, can only come into being when the light source has such a small angular size that it is almost invisible. And incidentally, this means that, applying this idea and forming interference fringes at the focal point of large instruments designed to observe the stars, using two broad slits quite far apart, we may perhaps expect to obtain more detailed information about the angular sizes of these heavenly bodies (Fizeau and Bordin 1868).

Fizeau justified his assertion by means of a simple calculation, treating the two "broad slits" as Young slits, and the starlight as a wave from the source illuminating them. And so a revolutionary idea was born. It would become the main tool in the quest to obtain sharper astronomical images right through the twentieth century. Note again Fizeau's modesty, presenting the idea as an incidental remark.

Shortly afterwards, in 1873, Fizeau's idea was put to the test by the astronomer Edouard Stephan at the *Observatoire de Marseille*, using a 1 m telescope, which was quite big for the time. With two closely spaced slits placed on the mirror, the fringes were quite visible to the observer's eye at the focal point of the telescope. However, the contrast did not fade, even when the two slits were placed as far apart as possible on either side of the mirror. Stephan gave the correct interpretation. Seen from the Earth, the star had a tiny apparent diameter. For the two slits which each received its light, this star was to all intents and purposes "a dimensionless geometric point", whence the presence of the highly contrasted fringes. Note that this method did provide a way to get round the notorious seeing barrier, gaining almost a factor of six, since Stephan was able to assert that the angular size of the star was less than 0.16 seconds of arc (Stephan 1873). But he was unable to actually evaluate the diameter because his mirror was too small, or, which comes to the same, because the angular size of the star was too small as viewed from the Earth. So these two Frenchmen came close to being the first in the world to directly reveal and measure the diameter of a star, other than the Sun, of course.

The First Nobel Prize in Physics Goes to an American

In 1891, Albert Abraham Michelson made another important step forward. In 1880, at the age of twenty-eight, following military studies, he did a tour of several European universities and visited the *Collège de France* in Paris. We do not know whether he actually met Hippolyte Fizeau, nor whether he even got to know of this remarkable idea about fringes produced in a telescope, which was only published in French. In 1881, whilst on a visit to Potsdam in Germany, not far from Berlin, he invented the incredibly simple device which now carries his name, the Michelson interferometer. This could be used to produce and study light interference in a great many different experimental setups, varying the path of the light, but also its colour, its polarisation, and so on. Just like Young's slits, the interferometer designed by Michelson had an extraordinary future ahead of it. Ever since then, generations of students have used this instrument to play around with light.

In particular, in 1887, it was used to carry out a very famous experiment, whose result would shake up the whole of physics. Thanks to a subtle and delicate observation, Michelson and his associate Edward Morley put an end to the hypothesis that there was an aether wind associated with the Earth's motion around the Sun. Ever since it had been recognised that light behaved as a wave, hence ever since the demonstration by Young, everyone had been wondering what medium was needed for such waves to propagate through. It certainly wasn't air, because it was clear that light could propagate through the vacuum of empty space. And so the aether was invented, a tenuous medium, unknown apart from the hypothesis that it was material and invisible, and that it could serve as a medium for the propagation of light. But the Michelson–Morley experiment showed that the aether didn't exist. It was time to think again about light, and in doing so, to reconsider space and time. The way was open for Albert Einstein's special theory of relativity, published in 1905. And regarding the interferometry designed for this experiment by Michelson, it too would have a glorious future since, in 2016, the first observation of a gravitational wave was made by the Laser Interferometer Gravitational-Wave Observatory (LIGO). In this case, the waves were produced by the coalescence of two black holes, something we shall discuss further in Chap. 7. The LIGO instrument was none other than a Michelson interferometer, but this time several kilometers long. What a magnificent fate for an idea as profound as it was simple!

Upon his return to the United States, Michelson pursued his interest in interference phenomena and applied it to measure the diameters of stars. Michelson's reasoning was a direct extension of Fizeau's, but let us emphasise once again that we don't know whether he was aware of this. We may only note that he didn't cite Fizeau in his first publication in 1890, nor in his paper in 1891, containing his first astronomical measurements. In any case, he opened the way to a very fruitful period that would last more than 40 years (Michelson 1891). Like Stephan before him, Michelson began with a rather small telescope (30 cm), at the Lick Observatory, not far from San Francisco, then moved on to a 1 m telescope, on which he set up a mask with two slits that would let the light through. He chose to observe Io, one of Jupiter's brightest natural satellites. This object has an apparent diameter close to 1 second of arc, as had already been established from an image taken when the seeing was good. He did indeed observe the reduced contrast of the fringes and was thus able to accurately measure Io's angular size. Fizeau's method had at last been applied with a positive result! But there were still the stars which, if they were to be resolved, would require a bigger mirror, bigger than the one Stephan had, and bigger than any available. At least, not for the moment. Then, in 1917, the

Hooker telescope was inaugurated on Mount Wilson above Los Angeles, with its 2.5 m mirror. However, even a distance of slightly more than 2 m between the two slits was still too small to diminish or wipe out the contrast in the fringes, so tiny was the apparent size of the stars. The slits would have to be moved even further apart.

With the help of Francis Pease, Michelson thus set up two mirrors, serving as slits to let the light through, at the two ends of a beam 7 m long, mounted at the top of the telescope. A further set of mirrors then sent the light to the main 2.5 m mirror, which formed the image of the fringes at the focal point of the telescope. This method could not remove the effects of agitation in the atmosphere, and of course Michelson did not have the benefit of adaptive optics. The fringes themselves were in constant movement, but slow enough to be able to observe them with the naked eye. Indeed, the eye can distinguish durations of the order of a tenth of a second. And in the same way, despite this agitation, the grape seed structure of a stellar image would later be observed by the Frenchman Jean Rösch and put to good use by Antoine Labeyrie, as explained in Chap. 3. At the time, no photographic plate was sensitive enough to pick up the phenomenon on such a short time scale, and the images observed by Michelson left no other trace than he described in the account he gave of his observations and measurements. This illustrates once again the vagaries of visual astronomical observation, whose most famous example must be the imagined canals present on the surface of the planet Mars, supposedly dug out by the Martians. Except that the fringes observed by Michelson were real enough.

Finally, 50 years after Fizeau's casual remark before the Academy of Sciences, Francis Pease and Albert Michelson observed the bright star Betelgeuse in the constellation of Orion. By an extraordinary feat of experimental ingenuity given the stability required for the telescope and its 6 m beam, the two physicists succeeded in making an accurate measurement of the apparent diameter of Betelgeuse. The excellent value they published is just one twentieth of a second of arc, that is, an angle twenty times smaller than the blur imposed by the seeing conditions. This was a major step forward. The stars were no longer just dimensionless geometrical points! The Dominican monk Giordano Bruno, so dramatically burned at the stake in Campo di Fiori in Rome in 1600 for his bold theological and philosophical stance, was the first, at the beginning of the seventeenth century, to assert that the stars were just like the Sun. Bruno argued that their appearance to the Earth-based observer as geometrical points was due to the fact that they were much further away from the Earth than the Sun. But that still had to be proved, and it was Michelson who finally achieved that.

Michelson died in 1931, just as his colleague Francis Pease was trying to improve on the previous measurement by tripling the distance between the two mirrors, setting the mirrors on a beam 50 ft (15 m) long. The measurement was not successful. The deflecting mirrors, which had been made bigger to gather more light, gave blurred images—they contained the famous speckles described by Newton and understood only much later by Antoine Labeyrie. Moreover, the vibrations of the beam completely swamped the interference fringes. Pease died in 1938, and the idea of coupling two distinct apertures was shelved for another 40 years.

Understanding Fizeau and Michelson

At risk of oversimplification, let me try to explain the basic ideas behind this important step in the war on blur, i.e., this key idea put forward by Fizeau and taken up shortly afterwards by Michelson. What is the origin of this crucial relationship between three quantities: the distance between the two slits, the level of contrast in the fringes, and the apparent diameter of the star that serves as the source of light, i.e., the tiny angle[6] subtended at the eye by this star when viewed from the Earth?

In the space surrounding a point light source, the emitted light is a wave that propagates in the three space directions, rather like the waves generated by a stone falling to the surface of a pond, except that those propagate in two dimensions. The propagating wrinkle is the wavefront. Far from the source where the observer encounters this wavefront, it looks locally like a straight line, like the swell on the ocean, generated far from the shore when it reaches the beach. The same is true of the 'wrinkle' of light produced by a distant star. Imagine for a moment that this star were actually reduced to a geometric point and place two holes, or two telescopes, to collect the light, at a certain distance from one another on the received wavefront. Now compare the vibrations of the light received by each hole. These vibrations will be identical—we say coherent—like the oscillations of the ocean swell when it arrives on the beach. Now increase the size of the star and move the holes further apart. The light emitted by different points on the surface of the star will not have followed exactly the same paths to reach the holes, and the coherence between the

---

[6]If a star like the Sun, with a diameter of 1.5 million kilometers, is placed next to the star Proxima Centauri, at a distance of 4.5 light-years (1 light-year $\approx 10^{16}$ m) from the Earth, the disk of that star viewed from the Earth will have an apparent diameter of $1.5 \times 10^9 / 4.5 \times 10^{16} \sim 3 \times 10^{-8}$ rad, or about 6 milliseconds of arc.

vibrations is reduced. But this means that the contrast in the fringes formed by interference between these two vibrations will also be reduced. The observer can thus conclude that, for this distance between the holes, he has shown that this star no longer appears as a geometrical point. The distance between the holes is then characteristic of the angular size of the disk presented by the distant star. This size can thus be measured.

The size of the Airy disk is a manifestation of the limit imposed by diffraction on the details of an image formed by a telescope, and this size is determined by the diameter of the telescope. The use of more widely spaced holes cannot remove this diffraction effect, which is always revealed by the presence of fringes, but it pushes back the limit imposed on the resolution. This limit is now fixed by the distance between the holes and the resolution is increased accordingly.

Note that Fizeau's powerful idea can be used to measure the size of a star if it is not too far away, but it cannot make a genuine image of the star's surface showing any of its details. Another important step remains to be taken. While the ability to resolve the details—in this case, the diameter—had been achieved by overcoming the blur due to the seeing effect, the production of a genuine image with this improved resolution was still not within reach of the telescope. The victory was incomplete, and yet it exploited principles that would one day prove to have much greater worth. The Nobel jury[7] who attributed the physics prize to Albert Michelson in 1907 "for his optical precision instruments and the spectroscopic and metrological investigations carried out with their aid", made no explicit reference to his interferometric measurement of the diameter of Io in 1891. And the diameter of a star had not yet been measured.

Albert Michelson's achievements in measuring the diameter of Betelgeuse and several other stars with Pease would be almost forgotten for the next half a century. When I was at high school in the 1950s, I heard about the experiment that had shown that there was no aether. This had made Michelson, and of course Morley, more famous than the measurement of Betelgeuse, which no one was talking about. Although we didn't actually repeat that experiment at school, we did use a Michelson interferometer, simplified but rich in all kinds of interference phenomena, under the attentive eye of Georges Guinier, an unforgettable and accomplished physics teacher. The common view was that Michelson was an experimenter of such extraordinary ability that scientists would never be able to do any better when it came to measuring the stars.

---

[7] www.nobelprize.org/prizes/physics/1907/michelson/biographical/.

Up until the 1960s, in the United States and Europe, several astronomers nevertheless made systematic measurements of the angular separations of a number of double stars that were closer together than the seeing disk. As Fizeau had done, they put two Young slits on their telescope mirror and thereby replaced the blur of the image by fringes fluctuating so rapidly that their contrast was only just accessible to the eye. These astronomers included the famous astrophysicist Karl Schwarzschild in Germany and also Maurice Hamy at the Paris Observatory. Although they could improve very slightly on the seeing limit, they were nowhere near producing true images with visible light!

Radio Waves Come to the Fore

In the meantime, and especially after the Second World War, a new branch of astronomy had been born. This was radioastronomy. And now, the relevant wavelengths of light were no longer a few ten thousandths of a milimeter as they had been for Fizeau and Michelson. They were measured in centimeters or even meters, which considerably reduces the requirements of instrumental accuracy and stability. Moreover, detectors sensitive to radio waves had been very widely developed for military purposes during the war, and in particular for radar in the UK. The physicist Yves Rocard had joined Free France in London. During the day, radar signals are perturbed by the Sun's presence. Yves Rocard thus turned his attention to the emission of radio waves by the Sun.[8] It was this wartime experience that led Rocard to set up the large radiotelescope at Nançay in Sologne, France, when he later became director of the physics faculty at the *École normale supérieure*, after the war.

At radiofrequencies, it thus seemed like a good idea to apply Fizeau's method, taking two samples of a radio wavefront with radiotelescopes several hundred or several thousand meters apart. This would avoid the very high levels of accuracy and stability required in the experiments by Michelson and Pease. In the 1950s, an array of radiotelescopes, purpose-built in Cambridge (UK) to carry out interferometry, began to reap the rewards of the exceptional sharpness it could achieve. The man behind this was Martin Ryle, who obtained the Nobel Prize in Physics in 1974 with Anthony Hewish, the first such prize to be awarded to astronomers. With a 5 km baseline—this term referred to the length of the beam, then to the separation of the telescopes in an interferometer—and an operating wavelength of 2 cm, this array of

[8] fr.wikipedia.org/wiki/Yves_Rocard.

radiotelescopes was able to attain a resolution of the order of one second of arc, hence comparable to the seeing effect limiting observations made with visible light. The radiotelescope array had an acuity five hundred times better than the diffraction-limited acuity that could be achieved by the mirror of a single radiotelescope of diameter 13 m. As far as the seeing or the agitation of the fringes are concerned, these present no problem for such radiotelescopes because, at the wavelengths used here, radio waves are almost completely unaffected by thermal fluctuations in the atmosphere! It's almost as though they were actually outside the atmosphere,[9] and their fringes are easily obtained and measured.

A True Image

Better still, Martin Ryle devised a method that would have a quite extraordinary future. In an array of eight telescopes, four were moveable, so it was possible to vary the separation between the telescopes, which thus played the role of the slits in Fizeau's conception of things. Ryle showed that a true image of the given astronomical source could then be built up from the interference between the light waves collected by each of the telescopes. Clearly, the blurring in this image was no longer the limit imposed by diffraction due to the mirror of a single telescope. Rather, it was determined, as it had been for Fizeau, by the longest baseline used. It was as though Ryle was using a huge mirror whose size was equal to the maximal separation between his telescopes, in this case, 5 km.

The only difference, although important, with an observation made using such a mirror concerns the amount of light collected. The total area of Ryle's eight mirrors was only a thousand square meters, whereas the corresponding giant virtual mirror with diameter 5 km would have an area of 75 square kilometers! This way of improving the resolution is known as aperture synthesis imaging. Although it does not increase the light-collecting area, this rather explicit term means that, with small pieces of a single giant mirror, one can synthesise the resolution of this giant mirror and even produce an image that actually resembles the object under study. The diffraction limit is then no longer determined by the size of these small pieces, but rather by the giant virtual mirror, which explains the enormous gain in resolution. The

---

[9]This is not strictly true. I just want to emphasise that radiofrequency interferometric observations were much more quickly and fruitfully developed than Fizeau's optical technique, because they are much less hindered by the presence of the Earth's atmosphere.

culmination of the ideas first developed by Martin Ryle can be witnessed in the Very Large Array (VLA), comprising 27 moveable radiotelescopes, inaugurated in New Mexico (USA) in 1980, followed by Europe's Atacama Large Millimeter Array (ALMA), which has been operating at millimeter and submillimeter radio wavelengths since 2011 at the Cerro Chajnantor site, not far from Cerro Paranal and the VLT, at an altitude of 5100 m, with 66 moveable telescopes. Another aspect of this tremendously successful development has been intercontinental radiofrequency interferometry (Very Long Baseline Interferometry, or VLBI), which I shall discuss further in Chap. 7.

The use of interferometry in radioastronomy thus began in the 1950s and flourished in the 1960s and 1970s, leading up in particular to the Very Large Array, and the interferometer set up to study the Sun in Nançay. During this period, many radio sources were discovered. Apart from the Sun, other stars, interstellar clouds, and galaxies also emit at the meter, centimeter, and millimeter wavelengths. Using the fringes measured by radio interferometry, images could be reconstructed, and I shall attempt to explain why shortly. These are indeed genuine images in the sense that they really do resemble the observed object. This thus went well beyond simply measuring a diameter or the separation of double stars, which was all that could be achieved at the time using optical interferometry. It was obvious therefore that it would inspire the bolder astronomers, dreaming of improving in one way or another on the achievements of Michelson at optical wavelengths, i.e., those of visible and infrared light.

From East to West

At this time, the Cold War was in full swing and, as I already mentioned when discussing adaptive optics, the US military would leave no stone unturned in their quest to increase the acuity of their satellite observations. Since the limit on the angular resolution[10] imposed by diffraction effects is inversely proportional to the wavelength of the light used, the transition from radiofrequencies to visible or infrared light can lead, for a given mirror, to a gain in acuity by a factor of more than a thousand! This was of course bound to attract some attention, and the self-interested curiosity of the Doge of Venice for Galileo's telescope in 1609 could be seen repeated by many officials at the Pentagon.

---

[10] In the present account, I sometimes use the correct scientific term 'angular resolution', and sometimes the term 'acuity', which suggests rather the ability to make out details. The reader will surely have noticed.

During the summer of 1967, a conference entitled *Synthetic Aperture Optics* was organised by the scientific advisory board of the Strategic Air Command, the unit of the US Air Force in charge of the nuclear strike force, parodied against an apocalyptic backdrop in the famous film *Dr. Strangelove.* The US National Academy of Sciences was also involved in the organisation and the conference was held at Woods Hole on the Massachussets coast, across the water from the island known as Martha's Vineyard where John Kennedy and his family had their summer residence. And so it was that fifty-seven people, including members of the military and a number of talented astronomers, discussed the possibility of repeating Michelson's master stroke. No European's were invited and the details of the discussions at this conference, considered top secret, were only publicly released much later. The published proceedings[11] open with a long quote taken from a book by Albert Michelson dating from 1903 and, incidentally, repeating almost word for word what Hippolyte Fizeau had said some 30 years earlier, but without citing his name. I mention this in passing!

However, the ball was soon to change court. In 1970, at the Pulkovo Observatory near the town of Leningrad in the Soviet Union (which returned to its original name of Saint Petersburg in 1992), Evgeni Stepanovich Kulagin went back to the ideas implemented by Michelson and improved them on an instrument built by the optical engineer Vladimir Pavlovich Linnik. This was the same Linnik who, 15 years earlier, had the idea of adaptive optics but, like Horace Babcock in California, had never made any serious attempt to develop it. Kulagin made an important innovation, replacing visual observation by photoelectric measurement. This provides an objective electronic recording of the interference fringes, more reliable than observing with the naked eye. This was indeed an important step, since it replaced a necessarily subjective assessment by an objective measurement that could be more easily communicated to and discussed by others. Kulagin studied the orbit of the bright double star Capella in the constellation of Auriga. He used a 6 m beam, as Michelson had done. His work was published in Russian but also translated into English (Kulagin 1970). Kulagin even filed a patent in 1963 with the State Committee for Inventions and Discoveries of the USSR.[12]

---

[11]Woods Hole Summer Study on Synthetic Aperture Optics, National Academy of Sciences, National Research Council, 1967.

[12]In 2018, Kulagin sent the text of this patent, in Russian and translated into French, to Daniel Bonneau, who kindly sent me a copy. Daniel, one of Antoine's first student, has written a remarkable book relating the interferometry story (Bonneau 2019).

Undaunted!

Recall that, in 1971, the young Frenchman Antoine Labeyrie observed the grape seed speckle effect with the Mount Palomar telescope, after first providing the theoretical explanation and announcing the potential it had for astronomical discovery. Recall also that observation of speckle allows one to go beyond the resolution limit imposed by the seeing effect, whence it is possible to make out finer details, some fifty times finer when a large telescope like the one at Mount Palomar is used at visible wavelengths. But Antoine was not satisfied with this result. He was already dreaming of imaging quasars, which would require still greater resolution of detail, and even black holes. And the same could be said of other astronomers, always keen to get more detail.

Naturally, there was no question of setting up a beam several meters long on the Hale telescope at Mount Palomar and trying to do things Michelson's way. The beam wouldn't even fit into the dome housing the telescope. The only solution was to use two independent telescopes, like the radioastronomers, because one can then adjust the separation at will, without the constraint imposed by a beam with its problems of rigidity, among other things. Antoine, this young man of just 30 years, didn't hesitate to take up the challenge, considering it well within his reach, and all the more so in that, in 1964, he had gone to see Vladimir Linnik and Evgeni Kulagin at the Pulkovo Observatory with his fellow students at Sup'Optique.[13] He had no doubt realised that using a beam to increase the distance between the two mirrors was a dead end.

In 1972, Antoine was still working at the Paris Observatory, housed in the beautiful campus on the great terrace of the *Château de Meudon* which looks down over Paris. He was doing all he could to promote his revolutionary ideas. He set up two small telescopes on the lawn, with manual, hence rather imprecise star tracking. The fringes remained elusive. His team included three young researchers, loyal partners and admirers.[14] They failed to obtain these fringes because the instrument was not stable enough and the atmospheric conditions in Paris too turbulent. Moreover, in those days marked also by a certain social turbulence, the gardeners at the observatory, prevented from mowing the lawn by this militant and sometimes less than diplomatic ecologist, saw no reason to be over-welcoming!

---

[13] The *Institut d'optique* Graduate School, nicknamed 'SupOptique'.

[14] These were Alain Blazit, Daniel Bonneau, and Laurent Koechlin. Many other young researchers were later equally keen to join Antoine in his adventures, but there wouldn't be room to mention them all here.

For my part, I had just been appointed professor at the newly founded University of Paris VII, a product of the Edgar Faure law which distributed the various faculties of the Sorbonne among a dozen specialised universities, while Paris VII was the exception because it was to be interdisciplinary. I joined a young team from Meudon working in the infrared and that was where I built the telescope that would be carried aboard the Caravelle aircraft, before going on to make an absolutely memorable observation of the 1973 solar eclipse aboard the prototype Concorde 001.[15] I thus found myself a good way from Antoine's work, which I could only admire from afar, and also from the ESO, about which I knew almost nothing.

In fact, it was precisely in 1972 that this new European organisation, set up on the CERN campus in Geneva, organised a meeting to examine instruments to equip the projected 3.6 m European telescope—no inkling of the VLT for the moment—, and other telescopes of similar size projected in France, the United Kingdom, and Germany, just as these countries were picking themselves up from the ruins of the Second World War. I have already discussed this ESO meeting in the last chapter. Antoine was invited because word had spread of his recent speckle observations and it was this work that he presented there. He rather boldly took advantage of this opportunity to talk about his longer term aim, so surprising to the audience that they had trouble taking him seriously:

> I suggest that future large telescopes, those currently under study, should be built in such a way that they can also be used as interferometers.

And rest assured, he was about to do it again.

Interference Fringes for Vega and Capella

Antoine set off on another brief visit to the United States. With his student Daniel Bonneau and others, he would make a series of observations using his speckle method at the 5 m Hale telescope on Mount Palomar between 1972 and 1974. Bright stars in the northern hemisphere, such as Aldebaran, Arcturus, Rigel, Betelgeuse, Antares, and Mira Ceti were measured more sharply than ever. This sometimes showed that these stars had an extended

---

[15] I have described this astronomical and aeronautic adventure in *Racing the Moon's Shadow with Concorde 001*, Springer, 2015.

atmosphere, something that could not previously have been seen on images with no detail.

When he returned to France, Antoine set himself up in Nice and, returning to his first failure in the grounds of the Meudon campus, he designed an improved instrument which he called the *Interféromètre à deux télescopes* (I2T). This was built on Mount Gros, the site of the Nice Observatory, which has a splendid view over the Baie des Anges. The observatory building was designed by Charles Garnier and the dome was built by Gustave Eiffel. The observatory itself was in the throes of the sometimes fruitful 1968 revolution. François Roddier and his wife Claude were teaching at the University of Nice and gave Antoine every support. François was an old schoolfriend of mine. We had shared the same enthusiasm for astronomy, then prepared the university teachers' examination together, in physics, for we both enjoyed the contact with students. François was astute, inventive, and an excellent and able physicist who loved carrying out experiments. He had a feeling for new ideas even before they had been properly implemented. Thanks to the mathematician Laurent Schwartz,[16] who had been our own teacher, François had acquired a thorough understanding of the tools invented by the mathematician Joseph Fourier, whose importance for image formation I shall discuss in some detail later on. François' role in the rest of this story will be discreet but important, sometimes even crucial.

In August 1974, the first fringes, obtained by Antoine on the bright star Vega, were photographed using a new photoelectric camera that was ultrasensitive to visible light (see Fig. 4.3). As in Pulkovo, the measurement of the interference fringes no longer depended on the subjectivity of the observer merely assessing the image with the naked eye. This was an objective quantitative measurement. The term 'optical interferometer' henceforth became the standard name for this kind of instrument, combining several telescopes—two in this case, and soon more than that. The distance between them is called the baseline, or baselines if more than one telescope is involved. In this case it was 12 m, which was more than had ever been achieved at Pulkovo in the Soviet Union, or indeed in the United States by Douglas Currie, another astronomer engaged in the same kind of investigation using a small interferometer with rather imprecise results.

---

[16]Laurent Schwartz (1915–2002) was a French mathematician who won the Fields Medal in 1950. He invented the theory of distributions, which play an important role in physics. In 1957, he was our unforgettable mathematics teacher at the *École normale supérieure*. He also distinguished himself by his radical opposition to France's war in Algeria. en.wikipedia.org/wiki/Laurent_Schwartz.

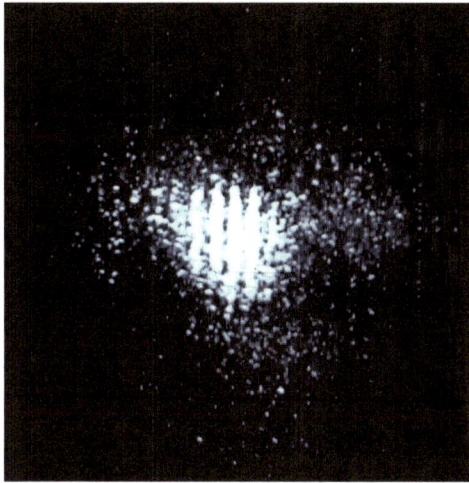

Fig. 4.3 Interference fringes covering the seeing-limited image of the star Vega in visible light, obtained by Antoine Labeyrie with the two-telescope interferometer (I2T) installed at the Nice Observatory. Credit: *Observatoire de Nice*/A. Labeyrie

The paper published by Antoine (Labeyrie 1975), this young man with so much self-assurance, in the Astrophysical Journal, the most prestigious American journal to which every astronomer dreams of contributing, is dated 1 March 1975. It struck the world of astronomy like a thunderbolt. It is said that, when informed over the phone, the renowned physicist and astronomer Robert Hanbury-Brown, who had put forward the idea in 1956 and had been investigating the possibility of setting up an extremely original but very different form of optical interferometry in Australia since 1963 (called intensity interferometry), immediately decided to drop that idea and build an instrument similar to Antoine's. And indeed the two met in Nice shortly afterwards.

Now, during this same period, astronomers in Europe, the United States, and the Soviet Union alike were puzzled by a particularly strange discovery. Radioastronomers had identified some peculiar objects which they called quasars, situated at great distances from the Earth but emitting phenomenal amounts of energy. Everyone at the international conferences was talking about these quasars, wondering what they could possibly be. They seemed to have particularly small angular dimensions, well out of reach of the acuity of existing telescopes. So it is easy to imagine that, after his first brilliant results, Antoine would be dreaming of one day determining the size of the brightest of these objects, 3C273, using his interferometer. And indeed, he often spoke to us about this. We shall return to this quasar and several others in Chap. 7, to describe how the VLT interferometer finally realised Antoine's dream in 2018.

The Calern Plateau, north of Grasse, is a vast limestone upland, at an altitude of about 1250 m, providing a magnificent karst landscape with sinkholes and limestone pavements. The nights are clear, far enough away from the light pollution of the night sky created by settlements along the French Riviera. Not far away on the horizon is Mount Mounier, from which a measurement of the speed of light, from there to Mount Gros in Nice, was made in 1903, using the method first tested by …Hippolyte Fizeau in 1849–1850. The Calern Plateau was selected in 1970 by the Paris Observatory to carry out laser–Moon ranging, in a project directed by the astronomer Jean Kovalevsky; specialist on celestial mechanics.

So what is laser–Moon ranging? The unmanned Soviet missions to the Moon that went by the name of Lunokhod, then the astronauts of NASA's Apollo XI, XIV, and XV missions, left reflectors on the surface. These so-called retroreflectors are made from three mirrors forming a trihedron with each face perpendicular to the other two—like the corner of a room, for example, and hence the name 'corner reflector'. The three mirrors of the corner reflector always reflect the light back in exactly the direction from which it came. Illuminated from the Earth by a laser light pulse, the corner reflector sends the light back precisely to the sender. By making a highly accurate measurement of the time taken for the return trip of the light pulse, slightly more than 2 s given the distance of the Moon, the position of the reflector can be determined to within a few centimeters relative to the Earth-based emitter. The motion of the Moon relative to the Earth can be deduced from this, better than it ever could before. In the limpid night of the Calern Plateau, a beautiful green beam of light would leave the Earth in the direction of our natural satellite, and I loved to take our students, still novice astronomers, to camp out on the plateau, a world away from Meudon. But very soon, and for a quite different reason, we would be sending other laser beams toward the sky.

Antoine moved from Nice to set up his interferometer with its two telescopes at the Calern site. The telescopes were placed on rails so that the baseline could be varied, eventually reaching a length of 67 m. When the baseline is long enough, the fringes begin to fade because the star is then resolved—at this juncture, the measurement shows that the star does not appear as a geometrical point to the observer on Earth. Its apparent size can be deduced immediately from this critical length at which the contrast in the fringes fades and disappears.

In 1977, Antoine's group published a second paper. It was no longer the distance between the two components A and B of the double star system

Capella that was measured, which Kulagin had already done at Pulkovo, improving on the measurements made in Michelson's day. It was indeed the diameter of each star that had been established, with a value of 5.2 milliseconds of arc for A and 4.0 milliseconds of arc for B. And although this result did not contain the wealth of information that would have been provided by a genuine image, the gain over the diffraction barrier corresponding to the telescopes of the time was quite spectacular, being a comfortable factor of ten! The reader would be surprised at the contrast between the rather do-it-yourself appearance of the instrument Antoine had put together and the revolution he had set in motion. New ideas often show up in this way!

With these observations of Vega and Capella, optical interferometry suddenly came back to life. By the end of the 1970s, astronomers the world over would flock to visit the Calern Plateau.

Take Courage and Make Your Point

In the last chapter, which tells the tale of adaptive optics, I already mentioned the important conference on telescopes of the future that was held in Geneva in 1977. Once again, Antoine would make some bold remarks, as he had done at the conference in 1972, cited a moment ago, but this time he could make his point with some assurance, given the interferometric results he had obtained in Nice and Calern. Using Michelson's method, he had successfully measured the stars Vega, then Capella, and using very small telescopes (25 cm), not even very widely separated (about 10 m). So on the basis of these results alone, he suggested that the future European telescope, still in limbo but quite possibly to be equipped by giant mirrors, of diameter around 10 m, hundreds of meters apart, should be set up as an interferometer!

Was This Reasonable?

Many in the audience were shaking their heads in disapproval. And yet Harry van der Laan, one of the future directors of the ESO, a radioastronomer well accustomed to radio interferometry, would make the cautious remark:

> I'm afraid I must abstain, through lack of time but not lack of admiration, from any direct comment on the progress of optical interferometry. I doubt whether the slight advantages that would be brought by a coherent array of optical telescopes could possibly justify the constraints and the costs.

However, an immensely talented optical engineer was present in the room. His name was Raymond Wilson. Of British origins, Ray, as he was known, had just joined the European Southern Observatory (ESO) where he was in charge of the design of the future and first large ESO telescope, with a diameter of 3.6 m, which was quite a respectable size at the time. This telescope would begin operation at La Silla (Chile) 2 years later. It belonged jointly to the then six member countries of the ESO: Germany, Belgium, Denmark, France, the Netherlands, and Sweden. Ray had spent 11 years in Oberkochen in Bavaria, working for the well known German company Carl Zeiss, famous for its binoculars and also the manufacture of large telescopes. But Ray's admiration was particularly drawn to the French school of optics, already often cited, and its modern offshoots. Later, when I had become properly acquainted with Ray, he would restate this admiration each time we met. He thus paid great attention to Antoine Labeyrie, even though he was fully aware that it would be a long time before such revolutionary ideas could be implemented for real in an operational instrument.

And indeed, it would be another 30 years. In any case, the seed had been sown and, 10 years later, Ray would be one of our most reliable supporters at the ESO, and later still, his own inventions and talents would be rewarded by prestigious international prizes, one of which would be presented at the French Academy of Sciences under the dome of the *Institut de France*. In the meantime, he invented a whole new design of telescope with a slightly flexible primary mirror whose surface could be corrected in real time for the effects of flexion and variable temperature. To avoid confusion with adaptive optics, we shall use the term 'active optics' here. Pushed through quickly under the impetus of Lo Woltjer at the ESO, funded with new money brought by two new member states, Italy and Switzerland, and inaugurated at La Silla in 1982, this New Technology Telescope (NTT) with its 3.58 m mirror also broke with the tradition of being housed under a dome, for this improved air circulation and thermal equilibrium as day turned to night. The concept of active optics demonstrated just what could be achieved by combining optics and computing (Wilson 2003), and it would turn out to be invaluable for the development of adaptive optics.

Following the 1977 conference, studies began to design the future European telescope and its most crucial component, the large glass primary mirror. Up to then, no industrial company, either European or American, had ever cast and polished a mirror of diameter greater than 5 m, which was the size of the extremely heavy mirror at Mount Palomar. What risk could reasonably be taken in going further? Cautious but keen to obtain a major gain for Europe over existing instruments, the director general of the ESO, Lo Woltjer, asked his engineers to study three options: a giant telescope with a 16 m mirror, four telescopes with 8 m mirrors, or sixteen telescopes with 4 m mirrors. The arithmetic was simple enough: in each of these cases, the same total amount of light would be collected, since the total surface area of the mirrors would be the same—around 200 m^2—although of course distributed rather differently. However, the risk of failure would be considerable for the manufacture of a 16 m mirror, somewhat less for an 8 m mirror, and almost non-existent for a 4 m mirror. So in the end it was down to the French and German companies to announce what they considered to be feasible without unreasonable risk. On the other hand, if there were going to be several telescopes, that meant that one could start thinking about an interferometer! Much later, when writing his memoirs (Woltjer 2006), the director general admitted that he considered this an interesting possibility, although he felt it would have been difficult to realise in the near future. And yet the VLT interferometer (VLTI) became operational at the top of Paranal just 3 years after the inauguration of the VLT!

In the last chapter I told how, after the rather disappointing reaction to the potential of adaptive optics and interferometry during the 1977 conference, we organised a further conference in 1981 at the ESO headquarters in Garching near Munich. This one was specifically aimed at these new prospects for improving image detail and the kind of astrophysical discoveries that could be made as a consequence.

In order to look a bit more deeply into these prospects, we invited Martin Rees. Although 10 years younger than me, he occupied the prestigious Plumian Professorship of Astronomy and Experimental Philosophy at the University of Cambridge (UK). This much coveted position had once been occupied by George Airy, the man with the blurred disk and the coloured rings, not to mention George Darwin and later, Arthur Eddington, the astronomer who used an observation of the 1919 solar eclipse to test one of the key predictions of the general theory of relativity proposed by Einstein. Needless to say, the opinions of the young Martin, still only in his forties, carried a certain weight! And since he was specialised in the remote and mysterious objects known as quasars that had only just been discovered, we asked him to tell us what further discoveries might be made in this field if we could some day overcome the blur.

I remember listening to him with a certain apprehension. Would he go along with our loosely sketched dreams, or would he shoot them down? As it turned out, although not at all well versed in the challenges of building observational tools, he had that visionary streak and supported our point of view. Given that the diffraction barrier imposed by the diameter of the telescope could indeed by reduced using interferometry, he even suggested that the improved resolutions might 1 day contribute to the study of black holes, which were still mysterious objects at the time. Decades later, I would often meet Martin, raised to the rank of Baron Rees of Ludlow in the county of Shropshire by Queen Elizabeth and appointed President of the Royal Society from 2005 to 2010.[17] The reader will meet him again in Chap. 7, when I explain how our investigations turned to black holes.

Among the three options put forward for the configuration of the future instrument, two of them could be used to do interferometry because they involved several distinct telescopes, either 8 or 4 m. Those attending this conference were thus faced with a new and fascinating question, although a highly controversial one. In addition to the challenge of equipping each telescope with adaptive optics, would it be possible to connect up several telescopes and make an interferometer? This would give the future optical observatory, already the biggest planned in the world, an acuity never before achieved on such a large scale. This was quite a programme! The two challenges together gave birth to the pursuit of what would become known as high angular resolution. This rather erudite term refers to the war on blur at visible and infrared wavelengths, with the aim of improving the detail in astronomical images.

What levels of performance could be expected from such an interferometer, built on a scale never before imagined? What major issues regarding the stars and galaxies—exoplanets were not yet in the news and black holes were only just coming under the spotlight—could such an instrument throw light on or even resolve? Would the engineers be able to handle the associated technical difficulties, never before encountered? And what about the budget? Would the European astronomical community accept to embark on such an adventure and assume the risks? In a word, was it really wise to take up such a challenge? These were some of the many questions we would have to answer over the coming decade, enough to strike fear into our hearts at times!

During the 1981 conference, a general discussion was organised the report it gave rise to is ambivalent, perhaps revealing the concern of some American colleagues not to let Europe get too far ahead (Angel 1981):

> There is a consensus, considering that a green light for the construction of a very large telescope is not recommended, even if funding was available. Indeed, there is clearly a need to put new techniques to the test, since they may lead to a spectacular reduction in costs.

---

[17] en.wikipedia.org/wiki/Royal_Societyanden.wikipedia.org/wiki/Martin_Rees.

We may understand here that it was not a good idea to attempt a 16 m mirror, which would be technically difficult. Small mirrors made using new techniques were preferable, but interferometry was not mentioned. It was anything but clear and this reflected the quite justifiable hesitation on the part of many of the participants.

We were ready to take the plunge, but we would still have to come up with sound arguments to persuade the powers that be. And time was running out! A decision would soon have to be made that would commit the ESO's member states to a certain choice for the main characteristics of the VLT. This would then be voted and construction would begin. It was 1981, and only six short years were left before this fateful moment.

The Idea Gains Ground

We moved cautiously. Jean-Pierre Swings is a Belgian astronomer who loves life, travelling, and what he refers to himself as 'astropolitics'. He was also a committed European like many of his fellow countrymen in the post-war period when the new European project was just getting up and running. Lo Woltjer, the director of the ESO, entrusted him with running the working group that would draw up the specifications of the future telescope and arrive laboriously at a suitable name: the Very Large Telescope, abbreviated to the VLT, a pleasing nickname by which it would soon become known to all. I was part of this working group with a dozen or so members, made up of ESO astronomers and engineers. From this point, I made frequent trips in Air France's noisy three-engined Boeing 727, back and forth between Paris and Bavaria, with its violent summer storms and its snowed up winter airport, its asparagus in spring and its goulash when the north wind blows, and its beer in every season—but above all, I went there for the VLT working group.

Our little group had regular meetings, and every idea was discussed in a surprisingly friendly atmosphere. At the same time, the engineers were analysing the three options with manufacturers who could cast and grind very large mirrors. These were essentially the German firm Schott and the French firm Reosc. It was quite soon realised that the 16 m option could not be realised using a 'monolithic' mirror, i.e., made from a single unit, because the risks of breakage were too great. At the same time, on the other side of the Atlantic, the famous Californian Institute of Technology (CalTech) in Pasadena, in the suburbs of Los Angeles, was considering the possibility of two giant telescopes with 10 m primary mirrors. They chose a different solution. The idea was to produce many small concave hexagonal mirrors, each different from all the others, then build these together very precisely into a mosaic in the form of a large concave mirror—this last was the delicate stage.

The team working on the future VLT preferred to leave the Americans to explore the mosaic solution, which was not without its difficulties, and in the meantime, the 16 m solution for the VLT was dropped. The European manufacturers said they were ready to make the mirrors needed for the 4 × 8 m option, without running too great a risk, while the least original 16 × 4 m option was also rejected. We would thus have four telescopes of diameter around 8 m. They would naturally be able to operate independently of one another, but also, from time to time, couple together to form an excellent interferometer, something we still needed a little more time to study in detail.

In 1984, Lo Woltjer set up a working group specifically to conduct this study and asked me to lead it. My instructions were precise:

> Set realistic scientific objectives for an interferometer using large telescopes in the optical range and also in the infrared, working out exactly what interferometry would impose on the specifications, and hence the cost, of a VLT.

The idea had thus reached maturity and would now officially take shape.

Obviously, Antoine was in the group, along with several other astronomers, including in particular Gerd Weigelt, who worked in Nuremberg and, as already discussed, had found a way to obtain genuine images from speckle observations. There were also some radioastronomers, whose experience with radiotelescopes in Cambridge and the Netherlands would prove invaluable, and three ESO engineers, including Fritz, the enthusiast already involved in our deliberations on adaptive optics, Ray, the master of optics mentioned above, and the chief optical engineer of the VLT project, the Frenchman Daniel Enard, who would soon become a close friend. All three were fully committed to the challenge of making an interferometer of exceptional acuity using the four telescopes.

Daniel was also a former student of the *École supérieure d'optique*, hence a spiritual heir to Fizeau and others in the French tradition. He was conscientious, level-headed, good-humoured, and thorough, convinced that we were onto something really interesting, but never complacent because in a certain sense he carried the whole success of the VLT on his shoulders and had therefore to weigh up every eventuality. Daniel left the ESO after the successful installation of the VLT. He would then go on to accept another, even greater challenge. This was the optical design of another kind of interferometer, although once again heir to Michelson's great idea. The instrument known as Virgo, set up not far from Pisa in Italy, was associated in 2016 with the historic first direct detection of gravitational waves, which was also the first direct proof of their existence, and probably the first direct proof of the existence of black holes, considered in this case to be their source. But alas! Daniel died before he could share in this second and most wonderful achievement.

Our meetings could sometimes be a little turbulent. Antoine brought to bear his imaginative style, always several decades ahead of his time, and I had of occasion to use a certain diplomacy to reconcile the various points of view. For example, Antoine considered—and he was right, on paper—that the interferometer would provide better images if the 8 m telescopes were moveable, like the radiotelescopes of the VLA. Daniel reacted strongly. I finally had to suggest, as a compromise, that a study be carried out by an Italian company, and they of course concluded that the costs and associated risks would be exorbitant: so that was the end of the 'telescopes on wheels'. This incident prompted the following irenic remark from Professor Ledoux, the imperturbable Belgian astronomer who chaired the ESO Council:

> Some have suggested [...] that it would be desirable to put all these telescopes on rails so that they could be moved around and used for interferometry, which would be a slightly risky undertaking.

A Giant Leap in Mirror Diameter

Our group had to deal with many different issues, but one of them was particularly decisive. We based our considerations on the modest results already obtained with interferometers, like the one built by Antoine, whose telescopes were of small diameter, only 10 cm or so. Just like Michelson's interferometer, these interferometers with small mirrors are not totally disabled by the effects of the Earth's atmosphere. Naturally, the interference fringes are moving quite fast, but they can still be picked up with sufficient accuracy by a camera on a very short exposure, of a few tenths of a second. But could these favourable results be extrapolated to the case where the primary mirrors, like those of the VLT, would measure almost 10 m?

Indeed, the situation here would be very different. The various parts of the wavefront reaching each of the large primary mirrors would lose some of their mutual coherence—what I referred to above as 'battered wavefronts' when discussing adaptive optics. When these seriously battered wavefronts are superposed to form fringes, the result of these lumps and bumps interfering with other lumps and bumps is an interference pattern covered with little disordered packets of fringes, scrambled therefore and hence practically unusable. Around

1982, with my friend François Roddier, we realised that this scrambling of the image would make it pointless to build an interferometer that combined the light from large telescopes, such as those of the VLT. Indeed, when observing faint astronomical sources, we need the great sensitivity of the large mirrors which gather plenty of light. If this gain over the expected performance of an interferometer operating with small telescopes were lost because the fringes were completely scrambled, what good would it be turning the VLT into an interferometer? So we published this rather annoying result, but we didn't completely give up.

As I explained in the last chapter, this was the time when adaptive optics was coming on the scene. If that could be developed for a large telescope, we were saved. The wavefronts would be straightened out, so to speak, before being combined to form interference fringes, so the scrambling effect would disappear and the fringes would then be perfectly usable. Therefore, if we wanted to be able to combine the light from the telescopes of the VLT to make an interferometer, we would simply have to equip each of them with adaptive optics. At that precise moment in time, such an idea remained something of a dream.[18] A few years on, when we wanted to get approval for the VLT interferometer, we would therefore need to show that this miraculous adaptive optics was actually possible, or indeed that it was already working. This was no minor challenge, but we have already seen how it was eventually achieved in Chap. 3.

From this point on, within our little European working group, we were convinced that we were following the right path. A little voice whispered in our ear that Europe, where modern science had been born four centuries before, could take back the lead from those upstarts across the Atlantic! If we really wanted to endow the future VLT with the exceptional level of acuity we were all dreaming of, we would have to open up a whole new territory, until then unexplored. We were just a handful of researchers to share this dream, spread out across Europe but supported at the ESO by some quite exceptional engineers.

In 1986, just as we were about to send in our conclusions, two colleagues in the United States, astronomers who were well known to us and whose abilities were above all doubts, published a paper in which they claimed that there was no hope of ever building an interferometer with large telescopes, precisely for

---

[18]In 1981, François Roddier published what would become a classic review, used by everyone, in which he discussed adaptive optics, apparently without much conviction at the time (Roddier 1981). He soon changed his mind, however, and would later make up for this, as we shall see in Chap. 5. Twenty years later he edited a magnificent book, *Adaptive Optics in Astronomy* (Roddier 2004).

the reason just given. Atmospheric turbulence would scramble the fringes and this would effectively cancel any advantage from the gain in light gathered by using large mirrors (Dyck and Kibblewhite 1986). We quickly responded to their argument during the presentation of the VLT project to the ESO Council in 1987. We just had to install adaptive optics on our telescopes and the problem would be solved! Their paper nevertheless contributed to some skepticism and criticism of the VLT interferometer project. Having said this, science moves forward in this way, because even dubious objections can be beneficial in some way and often lead to progress. So in the end, thanks to our colleagues across the Atlantic!

Making a Beautiful Image, from Fringes

Let us now see how an interferometer can produce a beautiful image, indeed a true image. How can one take the step from measuring the diameter of a star, as was done for Vega and the two components A and B of Capella using Antoine's small-scale interferometer, to producing an image that actually resembles these stars? This was the step that Martin Ryle had managed to take with his radiotelescope array in Cambridge just after the war, by developing aperture synthesis for radio waves. And since visible light, infrared light, and radio waves are all realisations of the same phenomenon, namely light, differing only by their wavelength, the ideas put into practice by Ryle would remain valid at the other wavelengths, even in the presence of unfavourable atmospheric effects.

My aim in the following, perhaps presumptuous, will be to describe the subtle physics of light and explain how a telescope can form an image, without using too many pictures or diagrams, even though the subject does get a little more complicated at this point.

What Is an Image?

The circular objective of a camera, usually a lens, forms an image of the observed scene on a light-sensitive film (in the past) or on the pixels of an electronic detector called a CCD (today) placed at its focal point. For our purposes here, a telescope is nothing other than a giant camera that works with a concave mirror instead of a lens. How can one then specify the size of the smallest details that can be made out in the image, i.e., what I have been calling the angular resolution to use the official term, or more suggestively and metaphorically, the acuity?

The angular resolution of the image formed at the focal point of the mirror is defined as the ratio of the wavelength of the light used and the diameter of the telescope.[19] This quantity represents what we have called the diffraction limit. It is the maximal acuity that can be obtain when the seeing is corrected by adaptive optics, or by placing the telescope in space. Adaptive optics had already shown us that, at a wavelength of 0.8 μm, which is in the near-infrared range, an 8 m telescope can achieve an acuity of 20 thousandths of a second of arc, which is fifty times better than the seeing limit.

The Legacy of Joseph Fourier

We naturally accept the familiar fact that a sound, like spoken words, for example, is just made up of acoustic vibrations of different frequencies, from low notes to high notes. A synthesiser can thus create complex sounds at will by superposing simple sounds, each with a different frequency and a different amplitude. In this way, the synthesiser can reproduce the timbre of any musical instrument, or indeed generate sounds that no existing instrument can produce. A complex sound can thus be represented as a sum of elementary sounds of different frequencies and amplitudes. This decomposition of a complex sound into elementary sounds has a parallel in the production of images from light waves.

Although it may seem surprising at first glance, an image can be understood as the result of the exact superposition of many regular straight fringes, alternating light and darkness, ranging from loosely to tightly spaced fringes, i.e., the exact analogue of acoustic frequencies, from the lowest to the highest. However, in contrast to sound, whose vibrations occur in time, hence in a single dimension, an image is deployed as we have already said on the two-dimensional surface of a film or a screen. Fringes are thus required in all directions on this plane to represent the details contained in the image. Note that the contrast of each set of fringes can also be adjusted, from the deepest waves to a tiny ripple.

Going back to the parallel between sound and image, a complex sound produced by an orchestra—a vibration changing as time goes by—is the super-position of many elementary sounds of different frequencies, from the lowest to the highest notes, each with a different amplitude. A complex image full of details—a distribution of light intensity over a two-dimensional surface—

---

[19]Put another way, the smallest discernible detail is measured by an angle $a = \lambda/D$, equal to the wavelength of the light divided by the diameter of the telescope.

is the superposition of many elementary systems of fringes, ranging from the most widely to the most tightly spaced, with every different orientation, each set of fringes having a different contrast.

By calculation, we can decompose a given image into its specific sets of fringes. Conversely, a given ensemble of these sets of fringes can be used to synthesise an image.[20] With the right collection of sets of fringes, each with the appropriate spacing, orientation, and contrast, any image can be obtained by superposition. This can be considered as analogous to the synthesis of a sound. But of course, the finer the detail that must be represented in the image, the more tightly spaced fringes must be used to achieve an adequate representation of those details. Our mobile phone bills increase if we ask our phones to transmit images that are rich in detail, with no perceptible blurring. Indeed, in this case the telephone signal must transmit more and more closely spaced fringes. This increases the amount of information, i.e., the number of megabytes (Mb) or gigabytes (Gb), that has to be transmitted.

This decomposition of a sound into sets of vibrations with different temporal frequencies, or of an image into sets of fringes with different spatial frequencies (spacings), is known as a Fourier decomposition. Joseph Fourier was a French mathematician who devised an extraordinary mathematical tool for representing these decompositions and calculating them. His work and its offshoots over the last two centuries now form the basis for all telecommunication of sound or image, and more generally for all signals carrying information.[21]

Aperture Synthesis

By collecting only two small samples from the light wave coming from the star using his two telescopes on rails, Antoine left aside a large part of the information carried by the wave. He thus obtained a single set of fringes, far from sufficient to form a beautiful image of the star Vega. When the baseline is short, these fringes show good contrast, because the star behaves as a geometrical point and is not resolved by the instrument. By increasing the distance between the mirrors, the fringes disappear. Why?

---

[20]There are many smartphone applications that can very easily carry out the Fourier analysis of a sound, revealing the spectrum of frequencies that make it up. A similar application exists for the analysis of two-dimensional images: https://2d-fourier-ios.soft112.com/.

[21]When the mainly science-oriented *Université Joseph-Fourier* was absorbed into the *Université Grenoble-Alpes* in 2016, this deprived Joseph Fourier, himself from Grenoble, of a well deserved homage.

A geometrical point is extremely small, in fact, infinitely small. To represent it correctly, we would need to have sets of fringes that were extremely closely spaced, in fact, infinitely closely spaced. But to reconstruct the disk of the star, which is not really just a point, we can do without such excessively closely spaced fringes. When the baseline is of the order of the wavelength of the light divided by the angular diameter of the star measured in radians,[22] the fringes disappear, because more closely spaced fringes would serve no purpose in representing the stellar disk.

What would be needed to go one step further and obtain a true image? The answer is quite simple. Since a pair of mirrors placed at a given separation (the baseline) and oriented in a given direction on the surface of the Earth provide a single set of fringes from the required image, we need only vary the distance between these mirrors (or telescopes), and also their orientation, in order to obtain every possible set of fringes one after the other, with every different spacing and in every direction. For each set of fringes, we then measure the contrast obtained between the brightest and darkest fringes—which astronomers call the visibility of the fringes, varying between 1 for perfectly contrasted fringes and 0 when there are no fringes. It remains only to combine all these sets of fringes in a computer and a true image will appear thanks to Fourier's miraculous decomposition, or recomposition in this case.

We can thus understand the advantage in having several telescopes available. Call these A, B, C, and D, supposing there are four of them, as in the case of the VLT. This gives the possibility of six pairs AB, AC, AD, BC, BD, and CD, which will not generally have the same orientation or baseline. We can therefore measure six sets of fringes at the same time, and thus six frequencies of the image. This is already better than just measuring one. If in addition the telescopes can be moved (telescopes on wheels), so that we can vary AB, BC, and so on, this is better still, because we can then explore the fringes of the image one after the other, from the least closely spaced to the most. The pair providing the longest baseline—say AD—is the one giving access to the finest details of the image, because it produces the most closely spaced fringes. The term 'aperture synthesis', already mentioned earlier when discussing radiotelescopes and their use in arrays, refers to this way of reconstructing an image using a whole sequence of these sets of fringes. The

---

[22]The radian, abbreviated to rad, is a unit of measure for arcs of circles or angles, introduced above. One radian is equal to $180/\pi = 57.3°$, or about $60°$. As viewed from the Earth, both the Sun and the Moon have angular diameters of $0.5°$, or 0.01 rad. The seeing limit is about 1 second of arc, which corresponds to 5 thousandths of a radian (5×10^{-6}). The angular size of the star Betelgeuse, viewed from the Earth, is twenty times smaller. At the end of this book, the tables of distance and angles review these values, which are given as examples.

principle is the same as for an acoustic synthesis. The word 'aperture' refers here to the diameter of the telescope that could provide the same level of detail. It is simply the length of the longest baseline, AD in this case, which is 130 m for the VLT. The term 'aperture synthesis' thus indicates that, instead of having a giant mirror, which is not feasible at the present time, i.e., a single 130 m telescope, which would obviously give incredibly sharp images, the interferometer reconstructs this image bit by bit, each bit being the result of a separate observation, so that the final result when all the pieces are put together will be equally sharp. It sounds like hard work, even somewhat miraculous, but it works perfectly thanks to Joseph Fourier.

We must therefore distinguish two quite different properties when an image is obtained by interferometric observation. The first is the level of detail it contains, determined by the most closely spaced fringes used to reconstruct it. The second is the resemblance with the object that produces it. This resemblance improves when more sets of fringes are used, from the least to the most closely spaced, and when these sets have all possible orientations in the image plane. The first property is determined by the length of the longest baseline, and the second by the number of distinct baselines used to carry out the aperture synthesis.

A Further Note on Speckle

Given what has just been described, let us return to the fact that the speckle or grape seed structure observed in short exposure images of stars contains information that can be recovered by suitable numerical processing. We are now in a position to understand this. Thinking the Fourier way, we now conceive of an image as a superposition of sets of fringes with every different orientation, and with the closest to the widest spacings—referred to in academic circles as 'spatial frequencies', by analogy with acoustic frequencies, which are temporal. When a light wave propagates in the atmosphere, the very small eddies it traverses cause plenty of 'damage' to the most closely spaced fringes, damping them and above all shifting them relative to one another—changing their relative phases—although without completely wiping them out. But these eddies will barely affect the less closely spaced fringes. So some information will nevertheless be gathered about the closely spaced fringes. Their superposition produces speckles. Computer processing can then recover this information and reconstruct some of the finer detail in the image, although not without certain limitations.

The Interferometer Takes Shape

With the long term objective of producing 'true' images by aperture synthesis, the VLT interferometer project was therefore designed according to this principle. This meant having as many different baselines as possible. In the version adopted by the ESO Council in 1987, the project would have four large telescopes, necessarily fixed, which could be paired together to provide six different baselines. There was much discussion between ourselves before we finally reached the best possible compromise in the choice of positions for the telescopes, on a mountain which was itself far from being chosen at the time.

Moreover, we suggested including two 'small' telescopes, referred to as auxiliary telescopes (AT), in addition to the four 'large' ones. Despite the name, they would each have a diameter of 1.8 m. The idea was that, being lighter and more easily moved around, this would considerably increase the number of possible baselines, the maximal length being fixed by the room available at the site. These 'small' telescopes would only ever be used for interferometry, even when the 'large' ones were making other observations, thus considerably increasing efficiency.

How would these auxiliary telescopes be made? The question was premature, but it almost led to an incident in our little working group. At the time, Antoine's team were building a new interferometer on the Calern Plateau, equipped with two 1.50 m mirrors (see Fig. 4.4). And so it was that the *Grand Interféromètre à deux Télescopes* (GI2T) would continue the work of the little I2T described above. These two telescopes were enclosed in two concrete spheres which could track the motion of the stars thanks to a novel mechanical system inspired by the way caterpillars move around. This was because Antoine is a nature-lover and keen advocate of biomimicry. Since we were proposing to add two smaller telescopes to the four telescopes of the VLT, Antoine wanted to introduce his spherical design, and not without certain sound arguments. But the ESO engineers quite rightly rejected any overly adventurous concept and the spherical design was abandoned.

This highly flexible combination of telescopes, including therefore four 8.2 m telescopes and the little 'egg-cups', would give the VLT interferometer its exceptional capacity to observe the black hole at the center of our galaxy, exoplanets in other stellar systems, and still more elusive objects, as we shall see later on, in Chaps. 6 and 7.

Fig. 4.4 On the Plateau de Calern (France), astronomer Antoine Labeyrie on top of the sphere, building one of the two *télescopes-boules* of the GI2T Project. His student Isabelle Bosc is standing below. Credit: D. Mourard

A Metro on Cerro Paranal

When the two telescopes constituting an interferometer receive light from a star, this light will not reach each of the two mirrors at precisely the same time, except in the exceptional case where the baseline is exactly perpendicular to the direction of the star. The difference in the time of arrival will be very small. For a baseline 100 m long and line of sight making an angle of 45° to the baseline, the path lengths differ by 70 m and light can travel this distance in about 0.2 microseconds. This time difference, called the phase difference, can considerably shift the interference fringes, and it can even make them disappear if the light is not strictly limited to a single wavelength. One way or another, this delay between the arrival of the light at the two telescopes must be corrected. The way this is done is simple enough, but rather cumbersome. Since the discrepancy corresponds to a path difference of about 70 m, the leading wave is given a suitable delay by requiring it to travel 70 m further using a set of mirrors, before combining it with the other wave. There is one more complication due to the rotation of the Earth, which means that the line of sight to the star—this angle of 45°—is constantly changing relative to the orientation of the baseline. The compensation for the delay between the two waves must be continually varied, although fortunately by an amount that is easy to work out, because the Earth's rate of rotation is extremely steady. It is imposed mechanically on the set of mirrors, which are carried on moveable trolleys, thus constituting the 'delay lines' of the interferometer, to use the technical jargon. The latter can be given the more musical name of 'optical trombones', since the mirrors can be made to slide rather delicately back and forth.

Our study group was well aware of the problem and the simplest solution, already used by existing small interferometers. We thus planned the construction of a tunnel 120 m long in which the temperature would be held constant, with rails on which a trolley could carry the set of mirrors. The motion of the trolley would be controlled by a computer to ensure that it was at the right position to create the right delay at any given time. Since the intention was to use up to four telescopes for interferometry at the same time, which meant six possible baselines, we would have to build six delay lines on six parallel tracks. We would rarely ever be able to visit the tunnel equipped with its long rails, and then only wearing special clothing and footwear to avoid introducing any dust. It would later be dug inside the Paranal mountain and referred to by some as 'the metro'!

Should We Take the Risk?

Our radioastronomer colleagues have already shown that such a combination of several telescopes could produce magnificent images. However, we had to be absolutely certain that we would be able to obtain similar results at optical wavelengths, which were a thousand or ten thousand times shorter, given that the Earth atmosphere causes much greater perturbation to the propagation of these wavelengths, and given the fact that we had far fewer telescopes than the twenty-seven dishes making up the Very Large Array in New Mexico. And it should be said that no interferometer operating at these wavelengths, either Antoine's or any of the other modest projects springing up around the world during the 1980s, had actually demonstrated the feasibility. Were we perhaps taking too great a risk?

We had long discussions about this in our group and I paid great attention to the various arguments. I couldn't help noting the confident attitude of the radioastronomers amongst us, especially Dennis Downes, a first rate astronomer who encouraged us to go ahead. Dennis was then working in Grenoble, at the headquarters of the *Institut de radioastronomie millimétrique* (IRAM), which was set up there in 1979. This was a joint venture between the

French government research organisation, the CNRS, and the German Max Planck Society with Spain associated, set up to run a millimeter wavelength interferometer on the Plateau de Bure, in the area called Dévoluy, located in the southern French Alps. The Joseph Fourier University in Grenoble was highly reputed for physics since it was developed by Louis Néel, winner of the Nobel Prize in Physics in 1970, and it happened to have an exceptionally active president at the time IRAM was set up, the physicist Michel Soutif (Schlenker 2016). In his view, astrophysics showed such promise in those years and was such an attractive subject for young students that he could not imagine leaving it out of the university curriculum. He had thus negotiated the installation of IRAM and then set about finding a suitable professor. Despite my love for the mountains, I was too well established in Meudon and turned down his offer.

In the end it was a brilliant physicist, a classmate of mine from the *École normale* called Alain Omont, who took the job in 1979 and thus moved to the Dauphiné region of France, where he quickly became an expert on millimeter astronomy. A group of young scientists gathered round him and they soon mastered radio-interferometry. A dedicated laboratory was set up on the campus, going by the name of the *Observatoire des sciences de l'université de Grenoble* (OSUG). It was only natural, then, that this expertise should extend to the kind of interferometry proposed at the VLT, when the basic ideas became better established. And this is indeed what happened. The *Observatoire* was soon full of young talent and grew quickly. Later on, it would play a central role in the development of infrared interferometry at Paranal, one of the main actors being Anne-Marie Lagrange.

Going back to speckle for a moment, I have already mentioned our young German colleague Gerd Weigelt, who was a member of our group. Astute by nature, he had that ability to push things forward with all those German qualities of rigour, perseverance, and a high level of organisation. Antoine was almost a god to Gerd, a visitation from a kind of opticians' heaven. We got on very well indeed, even though infrared light, my own speciality, was not 'real' light in his eyes! Only photons of visible light deserved his attention, and he handled those with remarkable ability.

Gerd would make a key contribution to the debate about the images, which was so important. He worked on the speckle patterns, as discussed in Chap. 3. Using the power of the fast computers that were just making their appearance in our laboratories, and exploiting a clever mathematical treatment inspired by the methods developed in radioastronomy, Gerd finally succeeded in reconstructing a true image despite the perturbation of the wave by the Earth atmosphere. The method, known as phase closure imaging, could be conveniently applied to the VLT interferometer. We could reasonably make this assumption and felt sure that time would prove us right. Indeed, a direct demonstration of phase closure imaging was given for visible light 9 years later by John Baldwin in Cambridge (UK), using his little three-telescope interferometer, the Cambridge Optical Aperture Synthesis Telescope (COAST). The VLT, although decided long before, was still not up and running. This demonstration came as no surprise, but it was nevertheless difficult to implement from a technical point of view, and it would have important repercussions, as we shall see in Chap. 5.

A Hard Choice: Infrared or Visible

By the end of the 1970s, we knew enough about the degradation of light waves when they pass through the Earth's atmosphere to understand that this degradation would be much less at near-infrared wavelengths. My previous trips to the Kitt Peak Observatory in Arizona, in the middle of the Sonoran Desert with its giant saguaro cacti, not far from the Mexican border, had put me in contact with some wonderful colleagues. In Tucson, one of these, Roger Angel, an outstanding optical engineer, had convinced the University of Arizona to set up a huge workshop for polishing telescope mirrors, underneath the no less immense football stadium. This was where Roger began to explore innovative mirror lightening techniques made from cast honeycomb glass. Roger's office was truly amazing! There were huge piles of papers, reports, and plans and the walls were covered with diagrams. The fact was that Roger had a perfectly original idea every 5 min, and would then spend hours checking it out with calculations and experimentation. Always smiling and welcoming, Roger impressed me enormously, and it was always a pleasure to discuss things with him. Very early on, even before the first studies for the ESO, we had reached the same conclusions, and had each published them. In short, we both felt that the future of optical interferometry was to exploit the properties of the Earth's atmosphere which strongly favoured the observation of infrared light (Woolf and Angel 1980).

So should we work in the visible or in the infrared? It was true that, from ancient times, astronomy had always been built up around the study of the visible light received from heavenly bodies. And although radioastronomy had by then really taken off, the same could not be said for infrared observation. This ancient study of visible light was therefore much more familiar to astronomers than the study of radio waves or infrared and sometimes tended to bias their judgement. Moreover, at this time, astronomers were gradually abandoning silver-based photography in favour of light detectors using electronic devices and exploiting the photoelectric effect in a solid material like silicon. These convert the light into electric charge and current. The first such devices were for the main part sensitive to the same wavelengths as those perceived by the human eye. In the case of visible light, in the range from red to blue, the images formed by such a camera show a tiny point of light on the screen for each impact by a photon. In this sense, we can genuinely 'see' the quantum nature of the interactions between light and the material forming the sensitive surface of the camera. It is relatively easy to analyse these impacts, just by counting them, since we are then effectively counting the photons. This principle is applied in charge coupled devices or CCDs, and it is what makes our portable phones into such good cameras when they are suitably equipped.

However, at that time, the cameras detecting infrared light that were just coming into existence were much less sensitive, unable to identify and isolate each photon impact. In this case, the measurement is not so 'clean', and harder to process. Infrared science is in fact a very recent subject. For Gerd as for Antoine, two eminent and much respected members of our little group, infrared light was not quite 'true' light. So what wavelength range should be chosen for the future interferometer: visible or infrared? Both Gerd and Antoine put all their weight behind the former, while the properties of the atmosphere and the question of accuracy spoke clearly for the latter. And yet these important considerations were not the only ones. Still more important was to determine in which of the two wavelength ranges the prospects for astronomical discovery were the most promising in the medium term, and with adequate supporting arguments. Because at the end of the day, only the strongest arguments would convince our fellow astronomers that they should approve interferometric applications at the VLT.

A Much-Debated Report

Our working group based its considerations on all the calculations and experience gleaned during the 1980s by those engaged in the war on blur, but also on the arguments put forward by our more skeptical opponents, whom we may refer to as 'interfero-skeptics'. The latter did not contest our calculations and observations, only the orientation we were advocating for the VLT and the arguments we put forward to justify it.

For my part, I had to reconcile all these different activities with my job as professor at the University of Paris VII, which was very important to me, the management of my research lab in Meudon, and my family life, with four children. It was no easy matter! There were moments of serious doubt. There were so few around us who seemed to share our hopes and it always worried me that I might be leading our young students up a blind alley. Could we justify the funding of this war on blur, to which I was asking my country to contribute? One person at least trusted in me and set me back on track: this was Steve Ridgway, the committed Francophile I have already mentioned.

In the beautiful autumn of 1985, I was walking with Steve in the pleasant gardens of the observatory in Meudon which looks over Paris. I explained my doubts and my reluctance to go on with the project, given the opposition from so many. Steve hardly hesitated for a moment before expressing, in his usual laconic way, his personal conviction that we were doing the right thing and that 1 day we would achieve our aims, even if it might take a certain time to do so. Often in the difficult times to come, I would remember this brief exchange and it did much to restore my confidence, such was the trust I placed in his opinion.

In 1986, after 2 years of work, we published our final report.[23] It concluded that the four telescopes of the VLT, with their 8 m primary mirrors, could indeed realistically operate together as an interferometer for at least part of their observation time. The prospects for astronomical discovery were considerable, and they could be achieved without disturbing plans for the other telescopes when they were used individually, which would be most of the time. We stressed that it was absolutely essential to use adaptive optics, and we felt confident that this would be feasible thanks to the construction of the prototype ComeOn, which was well under way. Our document provided a technical evaluation that confirmed the feasibility of the interferometry programme, and a financial assessment establishing a reasonable fraction of the total VLT budget to achieve the programme. There was then just one last step. The final report of the VLT project, which the ESO Council would use to take its decision, would have to validate our conclusion and include it in its overview of the project, its expected scientific benefits, and its cost.

What Could Be Achieved by Beating the Blur?

What galactic and extragalactic explorations should be undertaken? What questions should we try to answer? What astronomical discoveries could we expect? To obtain decisive answers, we would first have to make some

---

[23]Interferometric imaging with the Very Large Telescope, Final Report presented by the ESO/VLT Working Group on Interferometry, June 1986, ESO VLT Report No. 49.

reasonable assumptions about the performance of the interferometer we were recommending, and about the adaptive optics that would be essential to make it work: the resolution, final image quality, and sensitivity were the key features by which the project would be judged.

We naturally assumed that the resolution of each individual telescope would reach the diffraction limit associated with an 8 m mirror, viz., about 25 milliseconds of arc for the infrared wavelength of 1 μm, which is about forty times better than the average seeing disk! We could do ten or twenty times better still using an interferometer baseline of a hundred to two hundred meters to form a 'virtual telescope' with this aperture. A baseline of this length was about as much as could be hoped for on a rather narrow mountain peak like the one at Cerro Paranal.

However, it was harder to appreciate the quality of the final image, i.e., the resemblance to the object being imaged. This would depend on the number of telescopes, the number of baselines in simultaneous use, their relative positions, and hence the site chosen. There were thus many unknowns and the best we could do was to make certain assumptions and give a few examples, illustrating the most favourable cases. Much of what we said was based on the considerable experience of our radioastronomer colleagues.

At the end of the day, the greatest accessible sensitivity would determine the faintest objects that could be observed. Now, this sensitivity would depend critically on the quality of the light detectors available when the VLT was built, at least 10 years on from there. And during the 1980s, detectors were being improved tremendously, first for visible light, then subsequently in the infrared. We thus had to make reasonably optimistic assumptions. Time would show that what we had discussed in the report was a good assessment of the performance of the VLT interferometer when it finally became operational at the beginning of the 2000s.

Light from Other Worlds

In the 1980s, astronomers began to get a handle on the interstellar medium in our own galaxy and in others. This medium, usually rather cold (-250 to $-100°C$), is mainly composed of gaseous hydrogen in atomic (H) or molecular (H_2) form, although there are also small amounts of dust. The latter is made up of tiny silicate grains, either crystalline or amorphous, covered with water ice, ammonia ice, or dry ice, i.e., solid carbon dioxide. These dusts are formed in the atmospheres of ageing stars, then ejected into the surrounding space.

Stars arise by gravitational collapse from such clouds of gas and dust in the interstellar medium. An interstellar cloud, where the hydrogen gas is usually in molecular form, collapses under its own gravity. Eventually, one or more 'lumps' may collide and heat up as they fall, reaching a density that makes them opaque to their own radiation. At this point their temperature will increase even more quickly. The process continues until, in the end, a single large lump of hot matter is formed by accretion. If it is large enough, nuclear reactions may begin in the dense core and the object will light up, becoming a star.

At the same time as the formation of this stellar core or protostar, a disk of dust and gas about the size of the Solar System will form, rotating about the core. This disk is often associated with a flow of light-emitting gaseous material, ejected in both directions perpendicular to the disk. The evolution from a cloud to a star with this so-called bipolar outflow and surrounded by a disk takes place rather quickly, in less than a million years. At this time, a great many people were engaged in the study of ageing stars and the formation of young stars, in particular at the University of Arizona in Tucson, one of the most active centers for astronomy in the world, modelling these ejection and accretion processes and carrying out the associated calculations (Black and Matthews 1985).

One aim was to show that the protostar and the planets form more or less at the same time—in those days, such planets were not yet referred to as exoplanets. The idea seemed sound enough, but it was a largely theoretical result because few observations had yet been possible with a better resolution than the second of arc imposed by the seeing effect. An interstellar cloud located in the constellation of Taurus was particularly interesting because it was fairly close to the Sun, just 490 light-years away, and because it contained many very young objects that could be a sign of star formation. In the United States, using three telescopes with mirrors between 2 and 4 m located on Mauna Kea in Hawaii, four astronomers applied speckle interferometry in an infrared study. They ascertained that one of these objects, called HL Tauri, had a characteristic disk structure and jet of gas that could be observed on a scale of a fraction of a second of arc, hence smaller than the seeing. The paper presenting these observations had a title that held much promise: *Discovery of halos of solar system size around young stars* (Beckwith et al. 1984). One of its authors was Mel Dyck, a first rate astronomer I knew well, working at the University of Hawaii. Following our lead and courteously citing our work, he began to work on infrared speckle in 1980 (Dyck 1980), and he subsequently built a small interferometer. His opinion about large telescopes used as interferometers cited above clearly shows that we can all make mistakes.

In 1982, NASA launched the first space-based observatory specialising in infrared light, which was called the Infrared Space Observatory (IRAS). It would go on to make a host of exceptional discoveries.[24] Among the sources of infrared light that were detected, one was the environment of what are known as main sequence stars. These are stars one might describe as 'normal', in a long-term equilibrium with regard to their radiation, which is indeed the case of our own star, the Sun. In 1984, quite unexpectedly, a careful analysis of certain IRAS observations showed that the light received from these stars contained an excess of infrared radiation. This was attributed to the presence of a disk of cold dust and gas around the star (Aumann 1985). Could this be where planets were forming?

Let us look more closely at one of these, namely the disk around the star β Pictoris, in the southern constellation of Pictor. An inspired French astrophysicist, Alfred Vidal-Madjar, decided to devote all his energy to this— and he had plenty—surrounding himself as quickly as possible with a team of brilliant young students. He put forward a hypothesis which left many skeptical, that the disk in question contained solid objects like cometary nuclei which would fall from time to time onto the host star, just as happens in our own Solar System. Anne, whom we met right at the beginning of this story, joined Alfred's group in 1985 to do a doctorate under his supervision. Over the next three decades, she and her team would engage in a relentless effort to reduce the blur and thereby reveal the fascinating story of β Pic, with its disk and its first, but not necessarily last, exoplanet.

Our group at the ESO were looking for the best astronomical arguments to justify building an interferometer. The discovery of protoplanetary disks implied the development of a rich field of exploration of these until then unknown worlds (Bertout and Bouvier 1988). Moreover, the resolution of a few milliseconds of arc that could be obtained with the interferometer and the sensitivity provided by 8 m mirrors equipped with adaptive optics would be a guarantee of success.

Active Galaxies and Black Holes

During the 1970s, astronomers noticed that certain galaxies emit such a huge light output that there had to be some new underlying mechanism, quite different from the nuclear reactions which are the source of the light radiated by stars. It transpired that only the liberation of gravitational energy,

---

[24]en.wikipedia.org/wiki/IRAS.

transformed into kinetic then radiative energy when matter falls onto an extremely massive object, could provide a plausible explation for such high power outputs. This meant that, in the discussions we had been having about the design of the VLT since 1978, the possibilities for better observing these particular galaxies had become a central issue. It had already been estimated that, viewed from the Earth, the galactic nucleus where the light emission was generated had a tiny angular size, something like a millisecond of arc. It followed that the angular resolution that could be reached with the VLT would be crucial for studying such objects.

The galaxy Messier 77, also called NGC1068,[25] is only about 35 million light-years away from the Earth. This is much further than β Pic, but rather modest when compared to the distances of billions of other galaxies in the Universe. Messier 87 is an intense source of radio waves, but also infrared radiation, as was revealed by the satellite IRAS. Interpretation of these observations confirmed that we were dealing with a highly compact nucleus, and that the acuity of the VLT interferometer would be able to resolve it. One rather discreet member of the ESO with whom I particularly enjoyed working, the French astronomer Marie-Hélène Ulrich, a respected expert on active galaxies, was convinced that we were on the right track and gave us her support. This turned out to be invaluable, because although the resolution of the interferometer came well within the right range, there were some doubts about its sensitivity for such distant, indeed extragalactic, objects, i.e., situated outside our own Galaxy, the Milky Way. As a consequence, many astronomers who were quite rightly fascinated by cosmological questions considered that it would too costly and in the end futile to take the VLT in the direction we were suggesting. We were thus cautious in our predictions, although we knew full well that extragalactic astronomy would, in the decades to come, become a fascinating area of study if the VLT held its promise in beating the blur while reaching high sensitivity.

Protostars, exoplanets, and active galaxies were thus highly topical targets in our quest for sharp images. But these objects were not the only ones on the programme. Other more conventional subjects in which the benefits of improved acuity would be easier to reap were the surface of a star and its spots, the dusts it produced when it cooled, or the matter it ejected into space when it exploded in a supernova.

---

[25] A catalogue of relatively bright objects was made by the French astronomer Charles Messier in 1781. A more modern and much more comprehensive catalogue is the New General Catalogue (NGC), dating from 1888. The brightest objects thus have two different nomenclatures.

From the Report in Venice to the Decisive Council Meeting in 1987

The final report of these years of work was set down over a few wonderful days in the autumn of 1986 on an island in the middle of the Lagoon of Venice. About thirty of us gathered at the *Fondation Giorgio Cini* on the island of San Giorgio Maggiore. We had little time to contemplate the Madonna by Piero della Francesca. Our discussion focussed on the 300 page *Proposal for the construction of the 16 meter Very Large Telescope*, published in the spring of 1987 for the ESO Council, and hence also for the ministers of the eight member states who would have to approve the financing of the project, and of course for its future users, who would be astronomers from those countries.[26]

At the beginning of the 1980s, it was straightforward enough to draw up a long list of questions that could be tackled by the images and spectra obtained with a light collecting capacity equivalent to a 16 m mirror. Given the ratio of the areas, it would be able to collect $(16/5)^2 \approx 10$ times more light than the Mount Palomar telescope. This gain was already promising, and it wasn't the only advantage. The new electronic light detectors known as CCDs were twenty times more sensitive to visible light than the grain of a photographic film, and their extension to the infrared was already well under way. Taking these two things together, an object that would take an hour's exposure at Palomar would require only slightly more than a minute with the VLT!

While the observation of visible light and radiofrequencies was still the main occupation of most astronomers, Lo Woltjer quickly realised the tremendous potential of infrared observation and also the difficulties involved in its detection. At the beginning of the decade, he had even suggested that I join the ESO to take charge of the development of this new field. However, I was already busy in Paris with my university and family activities and couldn't accept. Alan Moorwood, a phlegmatic Briton with deadpan humour, whose immense talent left me in no doubt ever since we had once worked together to load a spectrograph aboard a NASA aircraft, became the main proponent of infrared astronomy at the ESO. He designed the first instruments for the telescopes at La Silla and made sure that this region of the light spectrum would be fully represented in the discussions in Venice. We shall find out more about Alan's legacy in the last chapter of this book.

---

[26]The story is told in detail by the chief engineer of the VLT project (Enard 1987).

In 1987, the document presenting the project to the ESO Council set the tone:

> At the present time, there are two clear problems that seem to be just beyond the reach of 4 m telescopes: one is the problem of the collapse of molecular clouds that results in the birth of new stars. The other concerns the birth of our own Universe and the study of distant galaxies.

This declaration referred explicitly to a remark by Edwin Hubble. A keen observer, Hubble concluded his book *The Realm of the Nebulae*, published in 1913, with the words:

> With increasing distance our knowledge fades and fades rapidly. Eventually we reach the dim boundary, the utmost limits of our telescope. There we measure shadows and we search among ghostly errors of measurement for landmarks that are scarcely more substantial. The search will continue. Not until the empirical resources are exhausted need we pass on to the dreamy realm of speculation.

Soon afterwards, applying these principles, Edwin Hubble would begin to exploit the potential of the new Hooker telescope on Mount Wilson (see Fig. 4.5). Sixteen years later, using this telescope, he discovered the

Fig. 4.5 The astronomer Edwin Hubble in 1923, seated behind the primary mirror of the Hooker 2.5 m telescope at Mount Wilson (California). Credit: Image available on Internet

cosmological expansion of the distances between galaxies. His successors would go on to build the 5 m telescope at Mount Palomar.

The conclusion of ESO document remains faithful to Hubble's way of thinking:

> The huge advance over present capabilities that the VLT will give will certainly lead to unexpected findings. Since nature is not constrained by the limitations of human imagination, we can be sure that the technical advances achieved by the VLT over smaller telescopes will lead to unforeseen discoveries which will revolutionize astronomy.

So what were these astrophysical questions that the VLT might be able to address? Let's not forget that we are still in 1987. Those other worlds, the exoplanets, were still effectively off the radar. However, the document went into some detail about the formation of protostars and the emergence of stars in the interstellar medium, forming from sparse clouds of dust and gas in the Galaxy. Its conclusion was as terse as it was cautious:

> These studies will be compared with information coming from Solar System studies to trace, for the first time, a complete history of the birth of a star. Only then will we have the empirical evidence needed to estimate the likelihood of other planetary systems or estimate the probability that other life forms exist in our galaxy.

The mention of black holes was equally brief and showed the same caution. These objects had been the subject of theoretical speculation since Einstein formulated his general theory of relativity in 1915, but their existence had been confirmed by the discovery of the Cygnus-X1 X-ray source in 1971. Sixteen years later, the document would propose a study of binary systems that were sources of X rays, noting that:

> It has often been supposed that the cataclysmic variable stars and galactic X-ray binaries may also yield insights into the phenomenon of black holes and the behavior of material around black holes.

With due consideration for the conclusions reached by our group, the final report devoted a whole chapter to adaptive optics for the VLT and another chapter to interferometry, closely relating these two proposed developments. It took a positive view about their prospects for success, considered to be likely. If we could at least go ahead with this in the infrared, the details of the images would improve by a factor of a hundred or a thousand, relative to the until then insurmountable seeing barrier. From this point, a whole section of the VLT programme was based on these innovations. It was now up to us to turn that into a reality, because according to the report:

> The Hubble Space Telescope will have neither the sensitivity in the infrared, nor the angular resolution, able to compete with the VLT.

What more could be said?

After the Venice meeting and during the following year, the project hit the headlines in the European and world press.[27] Not to be outdone, there was news from the world of astronomy, too, because in March 1986, the European probe Giotto photographed Halley's Comet from very close range, something never before achieved,[28] and then a year later, a star exploded in the Large Magellanic Cloud, becoming the first supernova of the twentieth century that could be seen with the naked eye.[29] In the French daily *Le Monde*, Jean-François Augereau wrote a piece under the headline "Eight European countries prepare to build the biggest telescope in the world".[30] Interferometry and adaptive optics had become credible options. In another French paper, *Le Matin*, Antoine made the enthusiastic claim:[31]

> Some day after the year 2000, we may be able to make out the street lights that illuminate the streets of extraterrestrial cities.

Preparing for Battle: Science and Astropolitics

It was now a year since the French Foreign Minister had appointed me a member of the ESO Council, the supreme decision-making body of the ESO, along with Jean-François Stuyck-Taillandier, a diplomat of great intelligence

---

[27] "Optical interferometry, made by modern technical developments, promises a new level of detail in the knowledge of celestial objects. Astronomers hasten to include it in plans for large new telescopes." (Thomsen 1987). The German weekly *Der Spiegel* paid homage to Antoine with the words: "French researcher develops a new kind of telescope", *Der Spiegel*, 17/11/1986. And the daily paper *Die Welt* noted: "Four mirrors increase astronomers' acuity", *Die Welt*, 20/3/1986. In the United Kingdom, the John Gribbin remarked cautiously: "Britain may also consider joining new projects with ESO in Chile", while quoting our friend Martin Rees, for whom "optical astronomy has a pivotal role for our study of the cosmos". John Gribbin, *Where next for astronomy and space research?*, New Scientist, 16 January 1986.

[28] en.wikipedia.org/wiki/Giotto_(spacecraft).

[29] en.wikipedia.org/wiki/SN_1987A.

[30] J.-F. Augereau, *Le Monde*, 16/11/1986.

[31] Robert Clarke, *La révolution des nouveaux télescopes*, *Le Matin*, 14/11/1986.

and equally great modesty (Stuyck-Taillandier 2012). We understood each other perfectly. After Venice, the next task was to ensure that France voted in favour at the next council meeting at the beginning of December 1987. Now it so happened that the political situation was complicated by the fact that the French president and the government were drawn from opposing parties. Since the legislative elections in 1986, which had been won by the conservative parties, the Prime Minister was Jacques Chirac, while the President of the Republic was still François Mitterrand, brought to power by a left-wing vote in 1981. Mitterrand's scientific advisor Jean-Daniel Lévi considered that the ESO's project should be approved, and he informed me of this. However, given the balance of power at the time, this decision lay outside the range of issues that came exclusively under the authority of the President, and a favourable decision from the Prime Minister was also needed, particularly regarding the significant long-term budgetary commitment required for the VLT. Indeed, this would necessitate an interministerial consensus.

The minister responsible for higher education and research at the time was Alain Devaquet, a scientist himself, who was hoping to maintain the impetus given to research since 1981. In December 1986, he was replaced by Jacques Valade following demonstrations in Paris that had led to the death of a young student, Malik Oussedine. A year of discussions ensued, during which budgetary questions often came to the fore. At the CNRS, Pierre Couturier, a future president of the Paris Observatory, was responsible for astronomy. We understood each other well and he often allowed me to have my say, ensuring that French astronomers would accept and defend the project. At the end of October, Germany asked to put the VLT on the agenda, and the research ministers of the two countries met. Naturally, the stakes were high concerning the manufacture of the large mirrors, in particular for the Germany company Schott and the French company Reosc. On 17 November, the minister Valade spoke to the Academy of Sciences in Paris, but his reaction was evasive: "If it were just up to me …".

Everything was now in the hands of the Prime Minister Jacques Chirac. From his office in Matignon, his new advisor on research issues called me and asked me to come and see him. We met quite late one evening. This high-ranking civil servant asked me to explain why the VLT was so important, adding that the Prime Minister was not yet convinced. I had no wish to develop any technical or financial arguments because the files available to him and the interministerial meetings he had attended would have been full of such things. I just spoke to him about astrophysics and the discoveries we would soon be making, and his interest seemed to grow in proportion to his ignorance of the vast universe. He asked questions and I replied, leaving his office at around nine o'clock in the evening: he would inform the Prime Minister of our conversation, he intimated as I crossed the threshold to go out. This last minute encounter was certainly a drop in the ocean compared with the preceding governmental machinations. But a few days later, Jean-Daniel Lévi called to tell me that everything had gone smoothly in Matignon. The French delegation could formulate a positive vote at the Council meeting on 5 December.

The Decision

The final decision to go ahead with the VLT was solemnly taken at this Council meeting. The ESO Director General and staff, who could not vote, sat with the sixteen representatives of the eight European states then members of the ESO. I had been there since 1986. Even though each delegation wanted a positive result, the discussions began to drag on. After all, it was a commitment to a project that would cost the equivalent of over half a billion German marks—this was before the euro—equivalent to about a billion euros today. No doubt recalling visits to Chile that had been affected by earthquakes, the Swiss delegate Peter Creola took the floor: "Gentlemen, you are about to make a historic decision [...]. Imagine for a moment your responsibility if, in a few seconds, an earthquake should interrupt our meeting before you were able to vote." To my great surprise, our arguments carried the day and were voted unanimously shortly afterwards.[32]

The decision made full allowance for what would become known as the VLT interferometric mode, or the Very Large Telescope Interferometer (VLTI). Almost unbelievably, we had won the first round, something quite unthinkable 10 years earlier.

Excitement and Doubts

Having described the internal processes at the ESO which set the scene for this historic moment in 1987, let us now go back a few years to take a closer look at the 1980s. During this decade, in Europe as in the United States, all kinds of ideas were being discussed and all sorts of prototype instruments were built. An

---

[32] Only one country, Denmark, held back its decision subject to agreement by others, but soon confirmed. The book *The Jewel on the Mountaintop* by Claus Madsen gives a lot more detail about the decision-making process (see the bibliography).

abundance of new observations came from these first small victories over the blur in astronomical images. These results were invaluable to us, because they helped to sort out the good leads from the bad ones, providing arguments that highlighted the potential for discovery of a very large interferometer. Moreover, they created many valuable personal contacts between people involved in this research, and motivated students to move into this new field. In this way, parallel projects sprang up, especially in the United States, but also in Europe.

The most extraordinary thing, and it is well worth emphasising here, is just the extent to which the decision in 1987 committing to the VLT was such a giant step relative to our knowledge and understanding at the time. A truly giant step and a gamble, too, as bold as they come, but nevertheless taken lucidly. The preparatory work accomplished during the first part of this decade was enough to show that the choice of building an interferometer was based on a sound scientific and technical examination. None of the remaining grey areas was extensive enough to endanger the project that was soon to follow. So let us take a closer look at some of this preparatory work which led to the choices that were finally taken on board.

Interferometers Galore

In 1986, the little Sydney University Stellar Interferometer (SUSI) set up in Australia after Hanbury-Brown's sudden change of strategy, inspired by Antoine's work, measured the diameter of Sirius, the brightest star in the sky, with remarkable accuracy. The Australian astronomer John Davis, whose careful work earned him the respect of the whole community, published a value of 5.63 thousandths of a second of arc. This result represented a considerable advance over the seeing barrier, obtaining a gain in image detail by a factor of 300 (Davis and Tango 1986). In the same year on the Calern Plateau, Antoine's group, combining interferometric imaging and spectroscopy, determined the size of the gaseous envelope surrounding the star γ Cassiopeiae as 3.9 thousandths of a second of arc (Thom et al. 1986). These results showed that interferometers equipped with small telescopes could already do interesting science, provided that the object in question was bright enough. We could only dream of what could be achieved with large mirrors!

Michael Shao, a brilliant young physicist of Chinese origin, was also inspired by Antoine's example to venture into interferometry. The huge Jet Propulsion Laboratory (JPL) where he worked was located in Pasadena, not far from

Hollywood.[33] In 1979, on Mount Wilson which looks over Los Angeles, Mike and his colleague Mark Colavita built small interferometers (Mark I, then III) which they used for astrometry, i.e., to measure the positions and displacements of stars, achieving accuracies that remained unrivalled for 20 years. Mike and Mark became good friends for us and invaluable scientific references.

Optical interferometers thus sprang up in Australia, California, and elsewhere. Our little club was beginning to convince the world that astronomy had a great future with such instruments.

From Berkeley to Calern

Young, Fizeau, Michelson, and Labeyrie had laid down a simple procedure for optical interferometry: take two samples of the wave front, superpose them, and detect the fringes. This was what the VLTI would be doing. Since Ryle, radioastronomers had been using a different method, better suited to the wavelengths they were studying. Instead of superposing the two wave fronts, they mixed each with a local radio wave source, which is easy to build at these frequencies. From the resulting mixtures and using well understood electronic techniques, they obtained the fringes and then the image by aperture synthesis.

Charles Townes, a professor at the University of California in Berkeley, had been exploring the interaction between radio waves and atoms and had built a wave source known as a maser. He obtained the Nobel Prize in Physics for this in 1964. The ideas brought to light here, which were essential to the development of lasers, invented shortly afterwards, also earned a Nobel Prize in Physics for the Frenchman Alfred Kastler in 1966. Charles Townes' area of expertise led him to radioastronomy and he detected all kinds of molecules in interstellar space. The presence of alcohols, ammonia, and formaldehyde revealed that there was quite complex chemistry going on out there.

Now it turns out that infrared light provides a good way to study such molecules, which may be signs of life elsewhere than on the Earth, and it can also be used to see the details of the clouds in which they are found. So why not develop an interferometer operating in the infrared using the light mixing method used by radioastronomers, with all its advantages? Indeed, in the mid-infrared there is a gas laser whose wavelength is characteristic of carbon

---

[33]This prestigious setup is a joint venture by NASA and the famous California Institute of Technology (CalTech), a private university. JPL has already had several triumphs in the exploration of the Solar System, including the Cassini–Huygens mission to Saturn.

dioxide (CO_2). It can be used to produce the local wave to be mixed with the wave from the sky. Using this idea, the team working with Charlie—as this rather stern gentleman allowed himself to be called—built an interferometer with first two, then three, moveable 1.65 m telescopes. Using this Infrared Spatial Interferometer (ISI) installed on Mount Wilson, the group successfully measured the diameters of many stars and also produced images of them.[34]

In the 1960s, Jean Gay set about exploring the infrared sky, just as I had. This creative young Frenchman, an able experimenter and imaginative physicist with an unstoppable determination, was keen to take Charlie on, exploring the same channels before him and independently of him. In 1978, his interferometer SOIRDETE ('summer evening') was set up on this same Calern Plateau where Antoine was himself building his own instruments at the time. The paper Jean published (Gay and Journet 1973) did not cite Charlie, but the latter, astonished to find this unexpected competition, showed great respect for his challenger. SOIRDETE did not have such a grand career. During the 1990s, slow to achieve concrete scientific results and subject to financial restrictions at a time when the VLT interferometer programme was absorbing a significant part of French resources, Jean's setup eventually had to cease operations.[35]

Convinced that his own approach was better, Charlie sometimes expressed a certain skepticism with regard to our strategy, while fully recognising that his own method could only work on very bright stars. The VLT could not be restricted to this narrow approach, but we often had to justify ourselves for not following the same method as the one recommended by a Nobel prizewinner! Maybe the two approaches will 1 day prove to be complementary once a successor to the VLT interferometer appears on the horizon.

Reinhard and the Black Hole

Berkeley, 1980. Charlie Townes has just entered our story. Dominating the beautiful bay of San Francisco, the Berkeley campus had fewer hippies now than during the 1970s when this university set the tone for student revolts the world over, rallying against the absurd Vietnam war. But Joan Baez could still be heard singing there. This campus remained, and still remains, a highly

---

[34]The interferometer ISI is described at http://www.isi.ssl.berkeley.edu/.

[35]Jean's two telescopes are still up there on the Calern Plateau. In 2019, this legacy of a heroic age is once again seeking out the stars. The subtle method of intensity interferometry invented by Hanbury-Brown in the 1960s was born again with very high performance detectors. And it could be that this method will prove fruitful in the future.

productive center for science, and a good place to do research. A young German with robust and determined features joined Charlie Townes' research group. His name was Reinhard Genzel. Since his doctorate in radioastronomy defended 2 years earlier, Reinhard had already acquired a sound reputation for excellence, something which had not escaped Charlie's notice. The following year, at the age of twenty-nine, the University of California appointed him as a fully tenured professor.

Another year on and the mysterious radiosource SgrA* was discovered at the center of our Galaxy. This was a fine subject of investigation for this brilliant young scientist who, in 1985, coauthored a paper with the prophetic title *Mass distribution in the galactic center*, with Charlie and a few others, in the illustrious British journal Nature (Crawford et al. 1985). This study analysed the motion of gas clouds in this central region, deduced from measurements at infrared and submillimeter wavelengths (the latter lie between radio and infrared). It concluded that there was in this region an unusual concentration of mass that was not in the form of stars, but was more probably a black hole of mass equal to four million times the mass of the Sun. Further investigation of this 'probably' would subsequently take up a good part of Reinhard's energy, until 30 years later, the whole idea had come very close to being confirmed by observations at the top of Cerro Paranal.

I met Reinhard for the first time at a conference in Cambridge in the United States, where, shortly after this publication, I presented the current state of our deliberations on the future VLT at the ESO. On the basis of his experience in radioastronomy, and having taken part in the interferometer programme that had occupied Charlie over the past 10 years in Berkeley, he presented us with an exceptionally clear analysis of the possible discoveries we might make if we had an interferometer operating in the near-infrared. I must admit that, after greeting his superb talk with the praise it deserved, I did feel slightly jealous. For Reinhard did indeed seem to have grasped the whole issue, even though he doubtless knew nothing of our careful analyses in Europe. Fortunately for European astronomy and for our future collaboration, Reinhard was ready to consider other options than the lure of California and chose to return to Germany. He came back as the director of the prestigious Max Planck Institute for Extraterrestrial Physics[36] near Munich, just a few 100 m from the ESO headquarters. This meant he could now bring his visionary lucidity to bear on the European interferometry programme which, as far as the VLT was concerned, was approved by the ESO Council the following year.

---

[36] https://www.mpe.mpg.de/main.

Giants on the Schedule Across the Atlantic

If we had been the only ones in the world to envisage giant telescopes operating in interferometric mode, we might well have been guilty of excessive arrogance. Fortunately, astronomers across the Atlantic were engaged in similar deliberations. It was essential to maintain a dialogue with them. From 1985, Lo Woltjer representing the ESO and John Jefferies representing the National Optical Astronomical Observatory (NOAO), a public institution supervising ground-based observatories in the United States, agreed to have regular meetings, sometimes in Europe and sometimes in the United States.

In 1947, the California Institute of Technology had inaugurated the 5 m optical telescope on Mount Palomar, and this had long remained the world leader. Caltech could not therefore remain inactive and began to prepare for a new telescope that would be financed by donations, in particular, from the oil billionaire W.M. Keck and a few others. The idea was to have two 10 m telescopes at an altitude of 4145 m at the top of Mauna Kea, on Big Island Hawaii, an excellent site where various telescopes were already up and running, including the Canada–France–Hawaii Telescope (CFHT). The construction of the Keck Telescope began in 1985, with a 5 year lead on the VLT. The two giants were placed 85 m apart, and its designers were well aware of their potential for interferometry. The latter, inspired by Mike Shao, planned to obtain fringes by combining their light.

Communication was rather limited, but I remember, around 1985, a meeting in California where we discussed the respective merits of the interferometers we wished to build. I presented our choice of supplementing the VLT with small auxiliary telescopes and went on to justify it. At this point, it seemed to me that there was a deep feeling of skepticism. So imagine my surprise when, 6 months later, I learnt that the Keck Interferometer, as it was called, would subsequently by equipped with five auxiliary telescopes, very similar to our own 'egg-cups'! In the end, only two of these actually materialised in Hawaii. On 8 May 1996, when the second of the two 10 m giants received its first light, a NASA statement announced the future interferometric coupling of the two giants, mentioning in particular the observation of exoplanets which had just been discovered at the *Observatoire de Saint-Michel* in Haute-Provence.

The W.M. Keck Telescope was the product of a private initiative and observing time was reserved for research groups at CalTech. So what was the strategy of the federal government of the United States with regard to future giant telescopes? In Tucson (Arizona), the NOAO had asked the astronomer Jacques Beckers to reflect upon the kind of telescope the United States should build, if they were not to leave all the expected discoveries to the VLT alone. Financed by federal public funds, it would succeed the modest 3.8 m Mayall telescope at Kitt Peak. Several options came under the spotlight for this future National New Technology Telescope (NNTT), which aimed at a diameter of 16 m, like the VLT. Jacques and his group very much liked the idea of making it into an interferometer, like the VLT. Jacques was originally from the Netherlands and had come to live and work in the United States. Specialised in solar observation, Jacques is a force of nature, an able experimenter, and a talented optical engineer, but a somewhat domineering boss (Beckers 2004). Once upon a time we had found ourselves in competition to measure the Sun's disk in the far-infrared. We thus knew each other. His group included one French member, my friend François Roddier, who had left Nice for Arizona. Alas for Jacques, the NNTT project encountered many setbacks. Finally, an international collaboration would lead to the construction of two giant 8.1 m telescopes called Gemini, one on top of Mauna Kea in Hawaii, the other on Cerro Pachon in Chile. Both were equipped with adaptive optics, but neither had interferometric options. But we shall meet Jacques again later …working for the VLT.

In 1987, we scheduled one of the annual encounters between Europeans and Americans at the Sun Space Ranch, close to a small town in Arizona called Oracle, a superb and perfectly authentic site for filming a western. The fifty or so people we invited to talk about the future of interferometry corresponded to almost every astronomer then involved in it, either in Europe or in the United States. On the recommendation of the director of the Shanghai Observatory, Yeh Shu-Hua,[37] I was even able to invite the young astronomer Qian Bo-Chen from China. This was just as the country was slowly recovering from the hardships of the Cultural Revolution. I first came into contact with the astonishing personality of Mrs Yeh on my first trip to the Middle Kingdom in 1979, which marked the resumption of scientific contacts between France and China, and we immediately established a friendly relationship. Protected by the Prime Minister Chou En-Lai, Mrs Yeh had escaped being sent off to labour in the fields, which was the fate of many Chinese intellectuals. She told me that, according to Prime Minister Chou, China could not survive without having access to exact time, and that only the skills of confirmed astronomers would be able to provide that.[38] Qian Bo-Chen left a rather distant legacy,

---

[37] en.m.wikipedia.org/wiki/Ye_Shuhua.

[38] Founded by the Jesuits in 1872, the Shanghai Observatory was an essential reference for this subject. Yeh told me she had spent the revolution with a pair of secateurs in her hand, since she was assigned to tending the roses in the park.

since in 2019, a group in Shanghai built an optical interferometer with three 10 cm telescopes.[39] Maybe in 20 years time, these astronomers will become absolutely essential partners in attempts to design a successor to the VLTI.

Discussions at the Sun Space Ranch soon revealed an increasing interest in the war on blur, and the following year, when the VLT had just been approved by the Council, a further meeting was held at Garching, under the heading *High-Resolution Imaging by Interferometry*. The two hundred or so participants, from 15 countries, were welcomed to the meeting by Fritz Merkle. The dense proceedings attest to the vitality of a research area that had only been around for a decade (Merkle 1998). For our own team, this meeting was tremendously encouraging.

Turbulence in France and Alliance Across the Rhine

Despite the fact that, in 1987, the decision to equip the VLT with an interferometric mode had been unanimously approved by the ESO member countries, it was clear that it would take 15 years or so to build and commission before any fringes could be obtained from a heavenly body. Everyone was getting worried about the loss of interferometric knowhow that might occur over such a long period. In France, Antoine's work sent out waves and he was soon elected to a chair at the *Collège de France*.[40] Thus, in 1986, we suggested that France could fund its own modest interferometer project, to be set up rather quickly as a kind of preparation for the ambitious VLT.

We started a study which quickly obtained Reinhard's support. A Franco-German interferometer project called Visir was proposed to the two research ministries on each side of the Rhine as a preparation for the VLT. To my surprise, the project was very quickly accepted, at least in principle. Michel Petit, director of astronomy at the CNRS, really wanted to see this strategy succeed, combining short and long term objectives! The physicist and crystallographer Hubert Curien was then the minister for research under François Mitterrand. Cooperation across the Rhine was a subject dear to this former member of the French Resistance, who had lived through and supported the construction of

---

[39] Optical Interferometry Laboratory of Shanghai Astronomical Observatory (OIL @ SHAO), english. shao.cas.cn/rh/rgs/201410/t20141009_128971.html.

[40] This choice was advocated by Pierre Joliot. With his sister Hélène, I believe these are the only human beings to descend directly from three Nobel prize winners: Marie Curie (who got two Nobel prizes), Frédéric Joliot-Curie, and Irène Joliot-Curie. On this occasion, Pierre Joliot described Antoine as a "crazy genius".

Europe.[41] No doubt the commitment of Reinhard and the Max Planck Society had a certain bearing on France's acceptance.

In parallel, we soon realised how keen the ESO was to get the now approved VLT interferometer up and running as quickly as possible, and it became clear that the Visir project was no longer really justified. We agreed that the budgets allocated to Visir on either side of the Rhine should be used to equip the VLT with an additional moveable auxiliary telescope ('egg-cup', or AT).[42] This agreement, signed in 1990, got the other member countries of the ESO moving and encouraged them a few years later to put up the money for a fourth AT. For example, the Netherlands would contribute thanks to the efficient impulse of Rudolf Le Poole. These now four moveable auxiliary telescopes would considerably increase the VLT's capabilities when forming images, since six interferometric baselines became available, corresponding to all possible pairwise couplings of these 'egg-cups'.

In 1985, Antoine set up his *Grand Interféromètre* on the Calern Plateau and he began to dream of another ambitious step forward. The Visir project seemed too modest to him. In 1988, just after approval of the VLT, he published his idea for the well-named Optical Very Large Array (OVLA), which he envisaged as an alternative to the VLT. Like the Very Large Array that had just been inaugurated by the radioastronomers, OVLA was to include 27 moveable ball-shaped optical telescopes arranged in a circle. While Antoine was quite right to stress the quality of the image reconstruction that an OVLA might perhaps allow with 351 baseline configurations,[43] he omitted to mention a more subtle, detrimental feature, related to the problem of sharing out the light between so many different telescopes.[44]

I tried for several months to work out a compromise in which the VLT would gradually integrate some of the features of OVLA and hence win over Antoine's support. The ESO engineers, led by Daniel Enard and Ray Wilson, who admired his amazing breakthroughs, could no longer understand his obstinacy. In the end I was forced to accept that my attempts were in vain. I

---

[41] https://en.wikipedia.org/wiki/Hubert_Curien.

[42] Two years later, on 24 December 1992, the journalist Catherine Vincent published a nice article about this agreement, by then operational, in the French daily paper *Le Monde*. It carried the title *La France et l'Allemagne s'associent autour du très grand télescope* and discussed the future of interferometry. A nice Christmas gift for us!

[43] Pairwise coupling of 27 telescopes can provide at most $(27 \times 26)/2 = 351$ possible baselines.

[44] It is certainly true that, as for radiofrequencies, the more telescopes and the more baselines there are, the better the image! But here one must physically share out the light and, quite the opposite of what happens with radiofrequencies, the more it is split up, the lower the sensitivity, and this quickly becomes catastrophic if one hopes to observe dim stars.

had to make every effort to ensure, from my position on the ESO Council, that the VLT interferometer would not become the victim of in-fighting between France and Germany, because we had opponents who would not hesitate to take advantage of the situation in Germany. Gerd was divided, ever loyal to Antoine but at the same time a faithful supporter of the VLT project for a decade now. In a very clear letter, in which I reiterated my respect for his work, I drew his attention to the disaster it would be if he withdrew from the project:

> The idea that the German community, of which you are one of the most eminent and respected members [in the field of interferometry], might withdraw from the project seems to me disastrous if it actually happens.

Gerd remained with us, while a few months later Antoine, a law unto himself, submitted his own project for the OVLA interferometer to the CNRS and the ESO. It was not followed up, as France was already fully committed to the VLT.[45]

Almost 20 years on, the VLT interferometer, which would not even exist without the basic principles first laid down by Antoine, is at present reaping in the rewards of all those efforts. We have both become members of the French Academy of Sciences and Antoine has since expressed a friendly respect for the rather difficult position I was forced to adopt at the time. Although it took time to come, this exchange was sincere, and it was one of the happiest moments of my professional life. OVLA never left the starting block. Antoine's *Grand Interféromètre à Deux Télescopes* on the Calern Plateau ceased all activity in 2005, without achieving many scientific results. Antoine turned to a revolutionary new telescope design which he is now testing in Haute-Ubaye in south-eastern France. Perhaps some of the ideas he used to advocate, and indeed still advocates, will provide a source of inspiration in the future, e.g., for an Earth-based optical interferometer that could supersede the VLT in the 2030s, or in a more remote future, for an interferometer in space or on the surface of the Moon. Because some day or other, humanity will want to obtain, with a new victory over blur, images of the surface of the exoplanet already discovered, so close to us at a distance of only four light-years, orbiting around the star Proxima Centauri, and of course images of many other such exoplanets.

---

[45] The author has the text of this proposal in his archives (4 September 1990).

Why Not Go into Space?

During the 1980s, NASA and the European Space Agency both clearly demonstrated their mastery of complex astronomical missions. Anything seemed possible. At the JPL, Mike Shao was considering the possibilities for an interferometer in orbit, or built on the surface of the Moon (Shao and Colavita 1992), which would be completely rid of problems due to atmospheric turbulence and the seeing problem. Antoine also dreamt of this. During the autumn of 1984, in the beautiful Corsican village of Cargèse, the European Space Agency organised a conference called *Kilometric Optical Arrays in Space*. Antoine presented a rather elaborate project which was called Trio, since it involved three telescopes moving together through space.[46] Trio was ahead of its time. Antoine's talks, then his lectures at the *Collège de France*, were often illustrated by a beautiful slide showing the Earth viewed from space, which he had deliberately slightly blurred, but still showing the green of the Amazon forest (see Fig. 8.3). This is what a planet like Earth would look like if situated about 30 light-years away and observed by an interferometer like Trio.

This was such a wonderful prospect that, during our discussions over the design and choice of the VLT, we could hardly avoid addressing the question. Why not prefer a spaceborne system to one that would inevitably be hampered by atmospheric agitation? We did indeed acknowledge the tremendous advantages of a spaceborne interferometer, but this still seemed a remote possibility and, according to the well known adage, "a bird in the hand is worth two in the bush". Fortunately, this argument was well received, because 40 years later, no spaceborne interferometer programme has yet seen the light of day. Mike and Mark would never realise their dream to place an interferometer in orbit around the Earth.

I always felt a certain sadness for our Californian friends.

---

[46]Proceedings of the Colloquium on Kilometric Optical Arrays in Space, 23–25 October 1984, Cargèse, Corsica, France, European Space Agency SP-226.

5

The Very Large Telescope: A Twofold Victory Over Blur

[…] the characteristic traits of grand passion—the immensity of the difficulty to be overcome and the black uncertainty of the result. Stendhal, *The Red and the Black*

Following the ESO Council of 1987, in which the VLT was approved by the then eight member states, Lodewijk Woltjer considered that this impressive success should mark the end of his heavy duties as Director General. He thus asked to be replaced. He was admired by all, and everyone was sad to see him go. Another Dutchman, the radioastronomer Harry van der Laan, was appointed as his successor for the next 5 years. I had known van der Laan for a long time. He was the one who would take on the difficult task of preparing the ESO for what lay ahead: construction of the VLT while respecting deadlines and budget, but also choosing the best site for it. He inherited a first rate team of engineers, whose numbers he would have to increase, while doubtless introducing further management procedures. Until then, under the friendly guidance of the optical engineers Daniel Enard and Ray Wilson, and also the Italian Massimo Tarenghi, the VLT team had been a tiny group with little formal structure that had designed the recently approved project.

The 1990s would not be without their difficulties, particularly concerning our war on blur, but nothing would keep us from victory in the end. It was going to be no mean feat to build the biggest telescope ensemble in the world, when the basic configuration was still to be determined and the instruments it would use had only just been invented, on a mountain that still hadn't actually

© Springer Nature Switzerland AG 2020
P. Léna, *Astronomy's Quest for Sharp Images*, Astronomers' Universe,
https://doi.org/10.1007/978-3-030-55811-6_5

been chosen, whilst remaining within the projected budget and maintaining the consensus of the eight member countries[1] that controlled the ESO. Of course, there were going to be obstacles, but 10 years after the approval in 1987, the ESO Council would meet at the Paranal site to inaugurate one of the most beautiful and powerful instruments in the world.

This particular decade, the 1990s, thus witnessed a magnificent adventure, whose heros were engineers and scientists, skilled technicians and accomplished administrators, and not forgetting politicians ready to lend an ear to astronomers in need of funding, and who in return could excite their imagination and contribute to the future development of the country, as I often heard tell. This makes a fascinating story, but it would be too long to recount here.[2] I shall thus restrict myself here to what concerns us most and what I was most closely involved in: the war on blur.

Choosing and Levelling a Mountaintop

Where should we build the VLT, and how should the set of telescopes be arranged? I shall discuss only a few of the more memorable episodes in this saga. The first ESO telescopes had been set up in Chile at the top of La Silla, 600 km north of the capital Santiago, since 1969. They were inaugurated by the Chilean president Eduardo Frei Montalva. His successor in 1970 was Salvador Allende, overthrown in a *coup d'état* in 1973 by General Pinochet. The La Silla mountain was a magnificent site that could accommodate the VLT. Not far from the peak was a plateau called Cerro Viscachas, which had all the right features. But might there not be an even better place, one where the weather conditions were even more clement, giving more clear nights? One where the atmosphere was less turbulent, with better seeing conditions, not to mention lower humidity in the air and hence better transmission of infrared light? Between 1985 and 1987, the first prospection attempts had come up with several relatively low-altitude summits (around 2700 m) along the Coastal Cordillera, not far from the Pacific coast, where the cold ocean currents, straight from the Antarctic, force cloud systems to remain very close to the ocean surface. This climate situation is what causes the Atacama Desert

---

[1]In 1987, the ESO had eight member states: Germany, Belgium, Denmark, France, Italy, the Netherlands, Sweden, and Switzerland. Others would later join, namely Austria, Spain, Finland, Poland, Portugal, the Czech Republic, and the United Kingdom, making 15 countries in all by 2019. Chile is the host country but is not a member of the ESO. Australia has a role as a partner country.

[2]I recommend the excellent account in C. Madsen, *The Jewel on the Mountaintop*, for more details (see the bibliography).

to be so dry, indeed, one of the driest deserts in the world. Further east, the peaks of the Main Cordillera, reaching heights above 6000 m, had to be ruled out due to their highly unstable weather conditions, something well known to Henri Guillaumet and the other heros of Aéropostale's Toulouse–Santiago link (de Saint Exupéry 1939).

It was while prospecting a site on the Coastal Cordillera that we almost had a tragic accident. At dawn on 30 July 1987, just at the time that government discussions in France for approval of the VLT project were under strain during the period of cohabitation between the socialist President François Mitterrand and the right-wing Prime Minister Jacques Chirac—as described earlier—, I had a phone call in Meudon from Lo Woltjer, who was still the ESO director at the time. His voice, ordinarily so self-assured, was trembling. Arne Ardeberg, a Swedish astronomer working at the ESO in Chile, had disappeared. He had gone off into the desert in a four-wheel drive with a chauffeur to explore the region around Paranal, but they had been caught in a violent snowstorm and the vehicle had broken down. Arne had separated from the chauffeur to go and get help on foot. But while the chauffeur had reached safety, Arne had disappeared. We knew he had a bag but no water and barely any food, and his footprints faded out somewhere in the desert. The following day, Daniel Hofstadt led the search party with sixty Chilean soldiers and a helicopter. However, Arne was nowhere to be seen. He made the headlines of all the newspapers in Santiago. Lo told me he was making hurried preparations to leave Bavaria for Chile. He reminded me in a weak voice that this desert had a particularly bad reputation. There was an internment camp for political prisoners of the current regime, and many of them would disappear in the desert. Much later, their wives would come back to look for their bones scattered among the stones of the desert, as told in the beautiful film *La Nostalgia de la Luz* by the Chilean director Patricio Guzmán. Arne was finally found safe and sound at the bottom of a *quebrada*, exhausted and thirsty, with no idea where next to turn.

As a precaution, the ESO had negotiated the allocation of an area of desert and the right to set up the VLT in Paranal. At the time, the Chilean government was under the control of the general and dictator Augusto Pinochet. This implied extraterritorial status, as in La Silla. During the 1990s, when Chile became a democracy once more, this allocation would be the subject of harsh criticism in the press and in the Chilean parliament, with the threat of legal proceedings by a Chilean family who claimed to have deeds to the land. Everything ended well, however, thanks to the skill and assurance of the then Minister of Foreign Affairs, José Miguel Insulza. Paranal would remain at the disposition of the ESO for 99 years.

The final report on the VLT, presented to the ESO Council in 1987, left open the choice between La Silla and Paranal, and recommended postponing the decision until 1990, when more detailed information could be made available, given that it was not yet urgent. Among the kind of information needed, some was absolutely essential in a country like Chile, where the Earth trembles almost continuously. I had great pleasure working with Philippe Bourlon, a friendly, imaginative engineer, who was afraid of nothing, lent to the ESO by the CNRS. The ESO needed him to study the seismicity of the two sites. Given that he would be studying the Earth tremors, he also accepted my proposal to measure the vibrations of the existing large telescopes in La Silla. Such vibrations would be highly problematic for the operation of our interferometer so it was important to learn as much as possible about the vibrational behaviour of very large metal structures. The conclusions, supplemented by a study by the *Institut de Physique du Globe* in Paris, were favourable. Years later, on 14 November 2007, an earthquake of magnitude 5.7 shook Paranal for about 2 min at midday, the strongest tremor recorded since the inauguration of the VLT 10 years earlier. There was no damage, and in the evening the interferometer was working normally.[3] Another tremor occurred one night while the interferometer was working. The fringes disappeared for a few moments, confirming that an interferometer makes an excellent seismograph, before reappearing again as if nothing had happened. This showed that all the necessary precautions had indeed been taken. However, quite apart from seismic activity, in 2019, the tiny residual vibrations of the 8.2 m telescopes remain one of the main sources of inaccuracy in the interferometric measurements, despite all the efforts by the engineers.

Those engaged in the war on blur soon made their choice, and by 1990 even those who remained more tentative had eventually opted for Paranal, the peak at an altitude of 2667 m, after reviewing all the material on the seeing quality, carefully collected by Marc Sarrazin over the years. However, there was no way to set up the four 8.2 m giants, given the topography of the mountaintop, let alone the moveable egg-cup array. The only solution was to level the mountaintop by taking 32 m off its height, thereby creating a platform that could accommodate an interferometer baseline 200 m long, as well as the necessary displacements of the egg-cups. The decision, put forward by the ESO management, was approved by the Council, and it was not long before the sound of dynamite came to disturb the silence of the Atacama Desert.

---

[3] See Earthquake News: http://www.eso.org/sci/facilities/paranal/news/previous.html.

The next problem was to decide how to arrange the large and small telescopes of the VLT on the Paranal platform. Interferometry itself would provide the arguments leading to the final choice. Indeed, our aim was still to obtain genuine images with the VLTI, and this would require fringes observed by pairs of telescopes forming baselines with different lengths and orientations. Since the 8.2 m telescopes would not be moveable, despite Antoine's earlier request, a trapezium configuration was the most favourable. It was chosen among other possibilities explored by numerical simulation. This choice was carried out in close collaboration with colleagues who had experience in radioastronomy, where the telescopes are moveable, as in the Very Large Array in New Mexico.

Things Could Still Go Wrong

The new Director General called upon one of his fellow countrymen to take charge of the interferometer project, namely, the astronomer Jacques Beckers, the 50-year-old Dutch astronomer encountered above. In the autumn of 1988, Jacques left Arizona and the now abandoned US project of a 16 m telescope. From there he travelled to the ESO in Bavaria, while, unfortunately for us, François Roddier, who had done so much to help our group from his base in the USA, chose to move to the University of Hawaii in Honolulu, where he would achieve the wonders in adaptive optics discussed in Chap. 3. To help him in his task, Jacques set up a new group within the ESO. One of its members was my young colleague Jean-Marie Mariotti, one of the rising stars of the new generation in France inspired by Antoine's work. Gerd and myself were in this group, and so was Antoine, until the disagreements I discussed earlier. Reinhard also joined us there. Apart from his foresight and his recent work on the still hypothetical but by now highly probable black hole at the center of the Galaxy, he had also become one of the powerful directors of the Max Planck Institute for Extraterrestrial Physics, with offices right next to the ESO. From my position on the Council, I tried to support the interferometer project and dispel certain prejudices regarding our war on blur. The Italian delegation was led by Franco Pacini, director of the Arcetri Observatory in Florence and the first astrophysicist to have imagined the existence of stars made up only of neutrons. He had been one of those who engineered his country's membership of the ESO. After all, how could anyone imagine a European astronomical undertaking without including Galileo's homeland? Their recent membership meant we had to proceed cautiously, and despite our excellent relationship, his delegation was not fully convinced by the interferometric perspectives, at least in the short term. To allow his group to deal with certain specific points, Jacques even set up what he called the Tiger Team at the ESO in 1990, a group of advisors whose name bears witness to the forceful vocabulary then in use among ESO management.

With the help of this tiger, which was not only on paper, but usefully produced plenty of paper, Jacques gave the interferometer project its final form. This laid out a detailed work plan for the following years, and was published a year after his arrival and 3 years after our initial report. The plan was unanimously approved by the ESO Council, despite the initial reluctance of the Italian representative. It would guide the work of the engineers and astronomers over the coming decade. The four 8.2 m giants would be set up on Cerro Paranal in a configuration minimising the volume of rock that would have to be removed. The elected trapezium configuration could give a long interferometer baseline, reaching 130 m with two of the giants and 200 m with the egg-cup auxiliary telescopes (AT), which was good news for the detail in the images! The tunnel referred to as the 'metro' was included on the plans along with the underground paths that would bring the light there from each of the telescopes. Since we had obtained permission to include two further ATs as an official part of the VLT project, later to be joined by two more, a network of rails would be laid out on the surface to be able to move them around. It would thus be possible to move the auxiliary telescopes over four different positions to be chosen among the thirty spread across the mountaintop platform, as the astronomers saw fit for each specific observation (see Fig. 5.1).

At the metro's exit, a large underground room was planned to house specialised instruments where the light from the telescopes could be combined. The interference fringes would be formed there, then measured. We spent a great deal of time discussing the size of this room, because it would have to be cut out of the solid granite of the mountain. The chosen area of 140 m^2 seemed enormous to our eyes. But 20 years on, the GRAVITY instrument, designed to observe the black hole at the center of our Galaxy, would make it look considerably less spacious when set up beside other fringe detection systems. Many optical and mechanical details which would be of great importance for the final result remained to be specified. Jacques, supported by the VLT's chief engineer Daniel Enard and his team, turned his mind to this with great talent, energy, and rigour.

One of the members of this team was the young French engineer Bertrand Koehler, a graduate of the *École Nationale Supérieure de l'Aéronautique et de l'Espace* (ENSAE). He had been seconded to our partnership as part of his national service 10 years earlier, where he had discovered the ESO and La Silla at the time when my young colleague Christian Perrier was using our

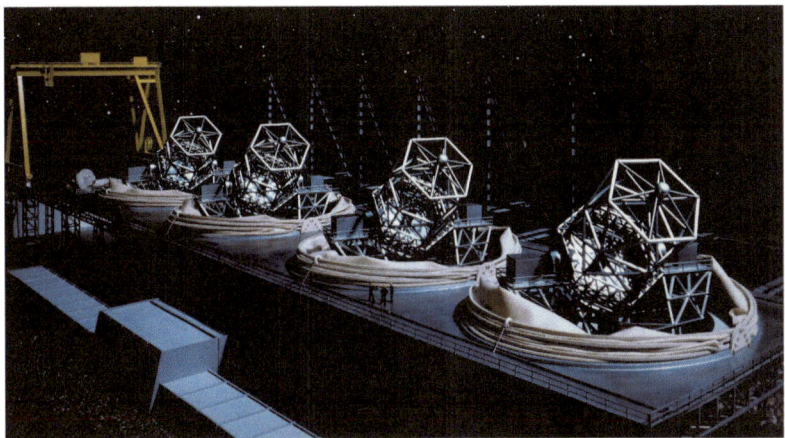

Fig. 5.1 *Upper*: A model of the VLT, as it was first imagined by the team of Daniel Enard and Massimo Tarenghi around 1985. The 8 m telescopes are aligned and retractable domes leave them entirely in the open air. In the foreground, the interferometric tunnel (the 'metro'). There is no platform to locate auxiliary interferometric telescopes (the 'egg-cups'). The star background has been added. *Lower*: The final configuration of the VLT, as decided in the early 1990s, and photographed after its construction 15 years later. The trapezoidal configuration of the four 'giants' is clear. Two of the 'egg-cups' are visible, as well as the rails on which they move. At center, the L-shaped building hosts the interferometric laboratory where fringes are formed from the light collected by each telescope and carried underground, where they are sent along the 'metro' with its delay lines. Credits: upper figure ESO and lower figure ESO/G. Hüdepohl

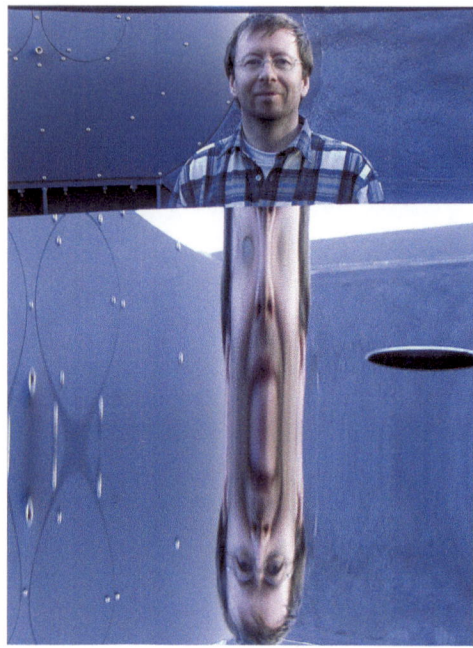

Fig. 5.2 Bertrand Koehler, ESO optical engineer and essential designer of the interferometer, looking at the surface of one of the VLT's mirrors. Credit: B. Koehler

infrared speckle imager (the *tavélographe*) to study the dust around stars, both young and old. Before joining the ESO, Bertrand worked for the company *Aérospatiale*, which was engaged in the Concorde supersonic plane programme. The company had a contract with Antoine, to build a delay line[4] for his small interferometer, and Bertrand was put in charge of it. At that time, the VLT had just been decided and Bertrand accepted the challenge and joined the ESO in Garching (see Fig. 5.2).

Working with Jacques in the small group of remarkable optical engineers gathered around Ray Wilson and Daniel Enard, he would then play the same role for interferometry as Fritz for adaptive optics. Thanks to his abilities as a systems engineer, many features were investigated, including the egg-cup auxiliary telescopes, the delay lines in the metro tunnel, and the effects of earth tremors already mentioned (Koehler et al. 2003). These years of detailed studies, carried out by Bertrand and his little team of French optical engineers

---

[4]This is what I referred to in Chap. 4 as an 'optical trombone'. Its role was to compensate for the continually changing differences in time, due to the Earth's rotation, taken by the light to travel from the star to the different telescopes making up the interferometer.

working at the ESO,[5] confirmed the validity of the initial choices. Some of these choices evolved with time, while costs were more accurately assessed and tenders could be prepared, in particular for European companies. As the months went by, our plans to build what had by now become known as the Very Large Telescope Interferometer (VLTI), although it still existed only on paper, seemed to go ahead smoothly without technical hitches.

But alas! The management of the VLT project by the Director General was criticised from the inside, and also by certain delegations on the Council, including the French, German, and Italian delegations, the three biggest financial contributors to the ESO, while the five smaller member countries remained more neutral. This broke an unwritten rule that we would always seek unanimity, as had been done under the previous mandate. In December 1992, the Director General's mandate was not renewed, and he was succeeded by Riccardo Giacconi in January 1993. Riccardo was an Italian astrophysicist who had, like many physicists of his generation, followed the example of the famous physicist Enrico Fermi and gone to the United States to pursue his career after the Second World War. His management style would thus be efficient and in line with the methods commonly practised across the Atlantic, something of a change for the European traditions at the ESO. After the end of his mandate, he shared the 2002 Nobel Prize in Physics for his discovery in the 1960s of X-ray sources in the Universe.[6]

As soon as he arrived at the ESO, he transformed the management procedures of the project and prepared the necessary budgetary measures, since the final cost of the VLT was estimated to go over by some 24%. Some of the resulting measures were painful. In the member countries, the VLT's rising costs had already forced reductions in national astronomical projects. I myself had to accept a difficult mission for the CNRS in 1992, one that was not so popular among my colleagues, advising the closure of certain observatories around France. Fortunately, by making suitable changes here and there, we managed to maintain the Pic-du-Midi Observatory south of Toulouse and the one in Saint-Michel in Haute-Provence, both the subjects of legitimate national and local pride. The first extrasolar planet was found at the latter in 1995, and this guaranteed its future.

At the end of 1993, a year after taking up his position and forced to take out a loan on behalf of the ESO in order to meet construction deadlines, Riccardo Giacconi put forward other measures for making savings. Among these was a

---

[5]Including in particular Samuel Lévêque and Philippe Gitton.
[6]https://en.wikipedia.org/wiki/Riccardo_Giacconi.

moratorium on the whole of our programme to combat blur, and indeed, he willingly confided his lack of enthusiasm for it. He qualified adaptive optics as a 'Christmas tree', while interferometry was just a costly dream in his view. To my mind, I could understand that several countries had similar views and would try to implement them at the ESO. But even the French government itself, whose scientists were the main instigators of this programme, ordered its delegation to vote in favour of this moratorium. I felt deeply disavowed by my colleagues and my country and considered that I could no longer represent it. Feeling powerless and heavy of heart, I handed in my resignation to the Minister of Foreign Affairs after this council meeting. Fortunately, I was replaced on the Council by Catherine Césarsky, a brilliant astrophysicist at the *Commissariat à l'énergie atomique* (CEA). Up to then, she had been involved in observation from space, and would now make the transition to Earth-based astronomy.

She soon got a taste for the new role, and by 1999, she succeeded Riccardo as Director General at the head of the ESO. For his part, Jacques Beckers left the ESO in the aftermath of this crisis. Even before, and despite the recommendations of Daniel Enard, he had not found appropriate to place the first industrial orders for the construction of the egg-cups. He went back to the United States, leaving the interferometer behind him. Its details had been pretty well worked out, but it had been shelved. He also left a young German astronomer he had taken on at the ESO, Oskar von der Lühe, an authority on high resolution observation applied to the study of the Sun. This crisis had its effect on the engineers and some of them left. Bertrand was in a quandary. Should he reject certain generous offers from institutions in the United States? He decided to stay, "if only to annoy the ESO by showing them in the end that the interferometer would work," as he told me much later, and he went on with the necessary studies.

For his part, the man in charge of the VLT project as a whole, the Italian Massimo Tarenghi, with his able Florentine disposition, who had helped us step by step to forge ahead with the VLTI since the 1980s, seemed relieved by the moratorium, given the complexity that interferometry would inevitably bring to the VLT project. Even though we may regret it, we can nevertheless understand his feeling of relief.

While waiting for better days, with scant resources, but supported by those few engineers whose creativity was stimulated by the challenge of interferometry, Oskar did everything he could to put together a rescue plan for the VLTI that might 1 day be put to the Council. A brilliant Italian astronomer, Francesco Paresce, who was at base somewhat skeptical about the future of interferometry, was appointed director of the now almost non-existent war on

Fig. 5.3 A visit to the SUSI Australian interferometer in Narrabri (1991). *From left to right*, John Davis, Australian astronomer and builder of the SUSI interferometer, with the author, then Daniel Enard, Head Engineer of the VLT Project at the ESO. The numerous optical components on the table recombine the beams from the telescopes, hence producing the interference fringes. Credit: Pierre Léna

blur. But suddenly, there was a glimmer of hope on the horizon, from far off Australia.[7] The prospect of joining the ESO would encourage the Australian commitment to interferometry, as mentioned earlier when we discussed the small but excellent Sydney University Stellar Interferometer (SUSI) run by John Davis (Fig. 5.3). Alas! At the end of 1995, the negotiations fell through and the hoped-for funds evaporated. So what could we now hope for? A change of heart by Riccardo Giacconi and the Council?

---

[7]This country is a partner in the British scientific tradition and has a very active community of astronomers. There is also an excellent optical telescope with a 4.2 m mirror on Mount Stromlo near the capital, Canberra. Australia would not have the resources to build a single giant telescope in the 8–0 m category.

Don't Give Up!

Coming after a period that had been so bright for me, those years were in many ways dark ones. But back in France, there was no way I was just going to give up. In 1994, the repercussions of the crisis at the ESO began to make themselves felt. At the CNRS, there was a suggestion that the French contribution to the third egg-cup might be postponed or even cancelled, even though we had already signed up for it. After the publication of an article in the British journal Nature which described the 1993 European disengagement from the Council, I stressed to our political leaders how much "others congratulate themselves on being able to take a leading position that Europe has abandoned, without even having to stoop".[8]

After all, if the VLT continued to suffer financial difficulties, it seemed possible that neither adaptive optics nor interferometry would ever be installed, or if they were, it would be in such a distant future that the whole strategy would have to be thought out again from scratch.

In France, stimulated by Antoine's creativity and the VLT project, there were more than a hundred astronomers and engineers in our research centers and in industry who were committed to this promising line of investigation and had no intention of dropping it. We had to consider the worst and imagine some kind of alternative plan. With Jean-Marie, and also with Antoine and his students, we set about organising the community of French astronomers engaged in this endeavour to improve the sharpness of astronomical images, while maintaining close links with Reinhard across the Rhine. The idea of setting up a national programme to work on high angular resolution (*Programme national sur la haute résolution angulaire*, PNHRA) had already been envisaged back in 1992. The CNRS asked me to take charge of it with the aim of aligning our efforts and studying the possibilities for such a plan. It would have to treat two issues: how to pursue adaptive optics, an area in which France was without doubt the world leader, and how to rapidly achieve scientific results with interferometry, which had initially been revived in France but was now being developed in an ever more competitive way around the world.

---

[8]Author's personal archives.

Adaptive Wonders

In Chap. 3, we left adaptive optics with the success of the instrument we called ComeOn, set up on the 3.60 m telescope in La Silla. During the 1990s, despite the frustration that had built up during this long crisis, scientific results were coming in all the time, and even Air France informed its passengers about this in the Air France Magazine.[9] The somewhat primitive prototype ComeOn was replaced from 1993 by the highly automated ADONIS, a system any astronomer could use without any particular expertise.

Norbert Hubin was one of those young French students attracted by astronomy. He began by signing up as an engineer in the service of national defence by joining the French aerospace research organisation (*Office national d'études et de recherches aérospatiales*, ONERA) during the 1980s. There he worked on the development of a laser weapon, something rather fashionable at the time, when people were talking about Star Wars. Working with industry, he learnt about the subtleties of atmospheric turbulence and its adaptive correction. In France, Renaud Foy had just published a surprising project with Antoine, for an artificial laser star, to which I shall return in Chap. 8. During the rather natural encounter between these two laser experts, Renaud told Norbert about the publication of a job offer at the ESO that would close the following morning. Norbert applied and was accepted, soon becoming an invaluable contact on matters concerning ADONIS, but above all the man who would ensure the future triumphs of adaptive optics in Paranal, having taken over from Fritz Merkle, who had since left the ESO.

The instrument ADONIS was equipped with two new cameras able to produce images abounding in pixels and sensitive across the whole of the near-infrared. The availability of these cameras attested to an improvement in international relations. In 1993, French astronomers were preparing an infrared camera for the European Infrared Space Observatory (ISO), whose launch in 1995 was imminent. This camera was the result of almost a decade of effort on my own part. It was to be put together in a partnership between the group led by Catherine Césarsky, still based at the *Commissariat à l'énergie atomique* (CEA) in Saclay, who were very experienced in matters of cryogenics,[10] and our own group in Meudon. We thus had access to an excellent image detector, a real gem from the technological point of view,

---

[9] Air France Magazine, 1995.

[10] These highly sensitive infrared detectors only work at very low temperatures, below −200 °C. Although it is essential, it is no easy matter to implement in space and requires specific technological know-how.

put together at the CEA's *Laboratoire InfraRouge* (LIR) in Grenoble. The atmosphere at the LIR was somewhat secretive owing to the military nature of some of their activities. The engineers and physicists working there were thus glad of the breath of heavenly inspiration brought by our astronomical projects, and they did everything they could to help us along. This same detector, with its 128 × 128 pixels, i.e., 16,384 pixels, a rather modest number compared with the tens of millions available in 2019, was developed for use in space, but also equipped ADONIS. In Germany, Reinhard had succeeded in importing an infrared detector from the US which had even more pixels. This detector, equipping a camera called SHARP2+, set it up on ADONIS in addition to our own. With him was a young researcher, Frank Eisenhauer, already mentioned at the beginning of this story on Paranal. There will be plenty more to say about him later on!

With this new instrument ADONIS, in La Silla and far removed from the ongoing crisis in Garching, interest in adaptive optics and its scientific results was growing all the time. By the end of 1995, almost half the available nights for observation with the La Silla telescope were allocated to those using ADONIS in the war on blur. Rather than just listing the rich harvest of sometimes surprising results (Léna 1998), let me just describe one of them in a little more detail, one that is particularly relevant to this story.

Recall that, in 1984, a beautiful image of a disk around a star had been obtained in visible light by two astronomers using a ground-based telescope in the United States. The disk was viewed edge on. Its dust scattered the light from the star Beta Pictoris and it looked as though it had been recently formed, like the star, perhaps about ten million years ago, by the mechanism described in Chap. 4. The star was hot, which made it more difficult to determine its distance than cooler stars like the Sun. Indeed, the latter have a spectrum that is richer in chemical elements and that helps us to work out the age of the star. For this reason, the distance (64 light-years) and age (8 million years) of Beta Pic—two quantities absolutely necessary to evaluate the properties of the disk—would not be precisely determined until 1997 (Crifo et al. 1997). Closer to the Sun than the Taurus molecular cloud, which was rich in newly forming stars, this disk and its details were much easier to analyse. It was quickly recognised to be a perfect laboratory for studying planet formation. Anne had just obtained her doctorate and her publications on the surroundings of Beta Pictoris were already piling up.

It was a beautiful December night in 1990, in the middle of the Chilean summer. In fact, it was New Year's Eve and I was at the controls of the instrument ComeOn in La Silla. With Anne in the rather cramped control room of the 3.6 m telescope, we were observing the star Beta Pictoris, surrounded by its disk. We were trying to reduce the blur with our new adaptive instrument placed at the telescope focus. At midnight, the champagne was flowing, and not just to celebrate the New Year, but also the deblurred image that had just appeared on the screen. The results were so convincing that Anne and I decided to build and integrate a coronagraph into the forthcoming improved ADONIS instrument. This device, the coronagraph, was invented long ago by the French astronomer Bernard Lyot to observe the solar corona. Using a mask, it would block out the bright light from the star Beta Pic which was dazzling our detectors, leaving only the faint details of the disk in the image. Two other students of ours joined up with Anne: Jean-Luc Beuzit to make the coronograph (Beuzit and Hubin 1993) and David Mouillet to implement it (Mouillet et al. 1997a).

We didn't have to wait long for the result, because the observations made using ADONIS, with an acuity ten times greater than the seeing limit, confirmed that the disk of gas and dust surrounding the star was indeed warped. This distortion could be interpreted as resulting from the gravitational attraction of a giant planet in orbit around the star (Mouillet et al. 1997b). This beautiful result was presented to the university jury by David Mouillet and he obtained his doctorate in astrophysics. So let us now await the commissioning of the VLT to see this planet directly in an image and pursue the story of Beta Pictoris and its disk in Chap. 6.

However important the results obtained with the ESO's 3.6 m telescope, we were not the only ones to explore this new field. And fortunately so. At the top of the extinct volcano Mauna Kea on the Big Island of Hawaii, the Canada–France–Hawaii Telescope (CFHT), built in partnership with a 3.6 m mirror like the one in La Silla, operates at an altitude of 4200 m, in a site with exceptional seeing.[11] It was inaugurated in 1979 and its first director was the Canadian René Racine. He, too, was committed, also quite alone, to the war on blur, taking a first step by stabilising image agitation and controlling the turbulence within the telescope. The ground was thus prepared for a first adaptive instrument. After some collaboration with Jacques Beckers in Arizona, mentioned just above, my long-standing friend François Roddier (see Fig. 5.4), supported by his wife Claude, herself an astronomer, came to live in Honolulu. He was joined there by François Rigaut, still revelling in the success of ComeOn, and the young doctoral student Olivier Lai, both leaving Meudon for the beauty of Hawaii. The adaptive optics instrument,

---

[11]See the bibliography, Thérèse Encrenaz, *La Montagne magique. Mauna Kea.*

Fig. 5.4 François Roddier, a long-standing friend indeed. In 1955, we were students together. I visited him as he was setting up his hand-made 20-cm telescope for the clear Burgundy nights. Photo by Pierre Léna

named 'Probing the Universe with Enhanced Optics' (PUEO), developed for the CFHT by this team, brought together some elegant new ideas,[12] and it soon demonstrated the relevance of these ideas by producing some magnificent images, also obtained in the infrared. The acronym PUEO was designed to coincide with the Hawaiian word 'pue'o', which means 'owl'. In 1996, their adaptive observation of the double star GG Tauri in this same molecular cloud revealed the presence of a young disk of diameter just 200 astronomical units, which meant that, in terms of size, it was of the same order of magnitude as the Solar System (Roddier et al. 1996). The disk, made up of a fine cold dust, was in orbit around two stars, and the magnificent image they obtained, where the seeing blur had been completely eliminated, was widely circulated.

The nocturnal vision of the new instrument PUEO, equipped with the latest infrared digital camera with a million pixels, made a long line of

---

[12]The ideas themselves are a little too complex to expose here. They were based on a new way of analyzing the distorted light wavefront, called a curvature analyzer.

discoveries right through to the 2010s (Lai et al. 1996). François and his wife Claude then took a well deserved retirement on the Mediterranean coast, after all their splendid contributions to astronomy.[13]

The originality and efficiency of this new adaptive design guaranteed it a fine future, in particular on the giant telescope Gemini Nord, with its diameter of 8.1 m, which received its first light in Mauna Kea in 1999, shortly after the VLT received its first light. Its anti-blur system was called Hokupa'a, meaning 'motionless', the name given to the Pole Star by the Hawaiians, who were excellent navigators on the Pacific.[14] François Rigaut was one of those who had contributed to our success that night in Provence back in 1989. Now established in Hawaii with his companion, the engineer Céline d'Orgeville, he built Hokupa'a, then set up a laser star on the 8.1 m telescope Gemini Sud. So many talented French scientists working around the world to beat the blur!

Regardless of the crisis at the ESO, adaptive optics was quietly making its way in the world, for it was by now quite clear that it would open up a tremendous field of exploration, particularly in the search for exoplanets, the first of which had just been discovered. Sadly though, we were forced to admit that we had no alternative plan should the VLT not be equipped with such a system, and sooner rather than later.

Would the Future of Interferometry Be Outside Europe?

In 1987, with the adoption of the VLT interferometry programme, Europe had given a clear indication of its scientific ambition and its intellectual lead. With the crisis of 1993 and with the long delays that seemed possible before the high resolution of the VLT came into being, what possible alternative plan could there be for optical interferometry in France and in Europe, given the growing competition across the Atlantic?

Indeed, in the United States, the Keck project was moving ahead quickly, combining the two 10 m telescopes to form an interferometer, and so was the construction of their auxiliary telescopes. To help the design of this coupling of the two giants, a new interferometer called the Palomar Testbed Interferometer (PTI), forming visible light fringes with two 40 cm mirrors, was set up by Mike

---

[13] François Roddier has since brought to bear all his creativity and originality on the study of the evolution of human society, considering it from a thermodynamic point of view: F. Roddier, *Thermodynamique de l'évolution: un essai de thermo-bio-sociologie*, éditions Parole, 2012 and *De la thermodynamique à l'économie: le tourbillon de la vie*, éditions Parole, 2018.

[14] www.spiedigitallibrary.org/conference-proceedingsof-spie/4007/1/First-light-for-Hokupaa-36-on-Gemini-North/10.1117/12.390369.full.

Shao at the foot of the 5 m Mount Palomar telescope in California (van Belle 1999).

A novel binocular telescope was under way in Arizona under the impetus of Roger Angel, who proposed to manufacture and polish the two 8.4 m mirrors in his underground mirror factory beneath the stands of the stadium at the University of Arizona in Tucson. These two mirrors would be attached to the same mounting and hence could be used interferometrically with a fixed baseline of 14.4 m. This instrument was to be called the Large Binocular Telescope (LBT), since it looked like a giant pair of binoculars whose mount would orient it toward the stars![15] This particular project made several appeals for European partnerships, at great risk of exhausting precious human resources for the VLT interferometer. Our Italian friends from Florence and those from the Max Planck Institute in Heidelberg could not resist the siren call from across the water and shared half the cost of the LBT.[16] Fortunately, given their weight on the Council, this choice did not dissuade them from supporting the rebirth of our own programme in the ESO when the time came.

Apart from giants like the Keck and the LBT, medium sized interferometers with well defined and promising scientific programmes were developing rapidly, like IOTA in Arizona and CHARA in California.[17] We collaborated with these two projects, but they still didn't really correspond to the genuine alternative that we had been hoping for and were having such difficulty in specifying.

For, in France, the crisis in the ESO and the quest for an alternative plan was complicating things considerably. With his group, Antoine had been running his *Grand Interféromètre à deux Télescopes* (GI2T) on the Calern Plateau for 10 years now. This group, including Daniel Bonneau, Farokh Vakili, and soon Denis Mourard, to mention but a few of the enthusiasts that Antoine had gathered around him, seriously lacked the funding to get GI2T running properly. It was not until 1997 that the first results were published (Mourard et al. 1997; Vakili et al. 1998). Antoine had built GI2T as a prototype for his OVLA project, which had disappeared from the French schedule and European deliberations after approval for the VLT.

---

[15] The two telescopes can operate as an interferometer by combining the light each receives. No delay lines are necessary since the baseline—the straight line joining the two mirrors—is always perpendicular to the direction of the observed star, so that the light from the star always arrives simultaneously at each of the mirrors.

[16] The LBT was inaugurated in 2004 and become operational in 2006: https://en.wikipedia.org/wiki/Large_Binocular_Telescope.

[17] I haven't included Charlie Townes' Infrared Spatial Interferometer (ISI) at Berkeley on this list. As indicated above, this used a different principle to detect the fringes, and was not in direct competition with the other instruments.

The crisis inevitably raised questions about this. In 1994, a meeting of the French national programme for high angular resolution (HRA) brought together some hundred and fifty scientists to take stock of the possibilities for an alternative plan.[18] The creative potential at that meeting was impressive. Should we drop our commitment to the VLT project, whose schedule had become so unpredictable, recover the further funding that the CNRS had planned to allocate to it, and turn to a French project like OVLA, which could be offered to European partners, using the spherical telescope technique of the GI2T? In itself, this was not without raising quite a few problems, just as the ESO engineers had anticipated long before. Alternatively, should we return to a French–German project like VISIR, worked out some 10 years previously? The often conflictual climate of uncertainty was sometimes hard to bear. I had no idea how things would work out, but then, in 1996, the clouds suddenly cleared. There would be no need for that alternative plan.

The future of interferometry had come so close to changing sides and crossing the Atlantic.

The Future in Europe Grows Clearer

In 1996, repeated efforts on both sides of the Rhine, and the studies pursued despite everything within the ESO itself, finally bore fruit. In June 1996, to get support from the European astronomical community, we organised a workshop with the ESO in which preparatory work was presented along with the scientific prospects for the VLT interferometer. The workshop was also attended by several colleagues from the United States whom we had invited to talk about developments across the Atlantic. The Swiss astronomer Michel Mayor was present and his support was invaluable. He had suddenly stepped into the limelight with his discovery a few months earlier of the first exoplanet 51 Peg b. Twenty-three years later, there would be more limelight when he would receive the Nobel prize in physics, along with his young collaborator Didier Queloz. Concerning the center of the Galaxy studied by Reinhard, the images taken with his SHARP camera revealed the motions of the stars around SgrA*, and there were more and more clues as to the presence of a black hole.

The ESO Council was finally convinced and decided at the end of June to revive the anti-blur programme (von der Lühe 1997). The Council thus

---

[18]VLTI: programmes astrophysiques, Programme National de Haute Résolution Angulaire, Lyon, 11 October 1994.

authorised the construction of an adaptive optics system and industrial manufacture of the various parts of the interferometer, including two egg-cup ATs,
with the third to follow soon after. Coupling of two out of the four giants
was scheduled for the year 2000.[19] This time, given the growing prospects for
scientific discovery, the French delegation led by Catherine gave its full support
to the decision, as did the German delegation, on the recommendation of Steve
Beckwith, who had discovered the disk around the star HL Tau while working
in the United States. Having returned to Germany, Steve chaired the ESO's
Scientific and Technical Committee. For his part, Riccardo Giacconi, who
was aiming for a second mandate as head of the ESO, doubtless considered
it would not be wise to alienate the French and Germans on this point,
since they had made their additional funding for the ESO dependent on the
decision to restart the project. Apart from anything else, he was certainly aware
of NASA's commitment to the Keck Telescope in the United States, which
was already building in an adaptive optics system and where the two 10 m
telescopes were on the point of being coupled together.[20] Before joining the
ESO, Riccardo had directed NASA's Space Science Institute and was fully
aware of the importance of this institution. All that no doubt contributed to
his reconsidering his previous rather negative point of view. Now in favour of
the interferometer and keen to hurry things along, the Director General agreed
to draw on a financial reserve intended for future VLT instruments (Fig. 5.5).

The new plan, presented by Riccardo and approved by the Council in
June, spelt out three top level research subjects that the VLTI was expected
to be able to throw light on thanks to the sharpness of its images. These
were the fundamental mechanisms of star and planet formation, access to the
immediate vicinity of SgrA* to "establish the presence or absence of a massive
black hole", but also access to active galactic nuclei, and finally, the search for
low mass stellar companions, including planets, in orbit around stars. For the
main part, the programmes identified in Venice in 1986, which had matured

---

[19]The first interferometric coupling of the telescopes was planned at the infrared wavelength of 10 μm.
Some of our German colleagues were interested in evolved stars and preferred these long infrared
wavelengths, while we in France were more concerned with young stars and disks, better accessible in
the near infrared, i.e., shorter wavelengths. The French were so unhappy about this initial choice that,
immediately after the Council meeting, Steve Ridgway suggested that we should instead invest the extra
money in the CHARA interferometer project on Mount Wilson in California. In the end, the whole
range of infrared wavelengths would be covered by the fringe detectors of the VLTI, i.e., the instruments
known as AMBER and MIDI, to be described below, in a well balanced solution for the institutes in both
countries, associated with a few others.

[20]In May, François Rigaut, stationed at the Canada–France–Hawaii Telescope in Hawaii, forwarded a
dispatch to me from NASA, which was "preparing to use the two giant Keck telescopes over the coming
years to search for planets and planetary systems around nearby stars".

Fig. 5.5 In 1995, the construction of the giant telescope structure progresses at the Ansaldo factory in Milan (Italy). The ESO Council members visit the installation of this 430 ton structure, which would later be moved to Paranal and receive the telescope mirrors. Some of the council members mentioned in this book are in the picture, including Jean-Pierre Swings (Belgium), Peter Creola (Switzerland), Riccardo Giacconi (ESO director general), and Franco Pacini (Italy). Credit: ESO

considerably over the past decade, were suddenly the main focus of the VLTI's future obligations.

We had a feeling that the three above themes would lead to a great many other discoveries. Everything was moving again. Optimism had returned to the ESO and our own research centers, but we would now have to work twice as hard. At the CNRS, Claudine Laurent worked with me to prepare French manufacturers for the ESO's imminent calls for tender. We knew that adaptive optics and interferometry would have to go hand in hand for the VLT. The first would ensure that each of the four large telescopes would overcome the seeing barrier individually. But the success of the first was also essential for the second to achieve its full potential. If we did not form fringes with waves that had been previously 'straightened out', it would be practically impossible to exploit the interference pattern produced by the giants, and the sensitivity to faint objects would be completely lost. It was crucial to adopt a coherent approach to these two aspects of the war on blur. This complementarity had been fully understood, so let us say more about it here.

The Rebirth of Adaptive Optics

Despite the crisis within the ESO in 1993, Norbert Hubin had done everything he could to prepare for the future of adaptive optics, for which he was responsible, given the limited resources at his disposal. In the end, the Director General agreed on a call for tender to European laboratories to build a single system that could equip one of the focal points of just one of the giants, Antu. Since the construction of an instrument of this kind involved many different skills and considerable funding, several French laboratories joined forces to answer the call. They formed a consortium which would thus be in competition with the project presented by the powerful Max Planck Institute for Astronomy in Heidelberg. Anne Lagrange would lead this French bid from the campus of the Joseph Fourier University in Grenoble. Under her leadership, the group of laboratories forming the consortium, including our own team in Meudon represented by François Lacombe, well prepared by their experience acquired with ComeOn and ADONIS, made the winning bid.[21]

The instrument would be called NAOS, an abbreviation for Nasmyth Adaptive Optics System. Since NAOS had to analyse and correct atmospheric blurring in the images, it would be associated with an infrared camera. As a kind of consolation perhaps, the construction of the CONICA camera would

---

[21]But only just! Indeed, one member of the consortium was of course the *Office national de recherches aérospatiales* (ONERA). This French institution depends on the Ministry of Defence. It had become interested in adaptive optics from very early on and had been our key partner in the successful project of 1989. However, the financial rules they had to abide by meant they had to bill each hour worked at a level the budget allocated by the ESO for the project could not cover. Our European competitors had no such restriction and it looked as though they might take the contract if there was no way we could lower our costs. Denis Maugars, himself from ONERA, was a member of the cabinet of the young Minister of Research, François Fillon. He immediately realised what was at stake and managed to raise the required amount. Our proposal would thus win the day!

be attributed to our colleagues and competitors in Heidelberg. This pair of instruments, eventually set up at the Cassegrain focus[22] of the giant telescope Yepun and given the name NaCo (NAOS-Conica), collected its first light at the beginning of the year 2000. As soon as it began to operate, NaCo had one success after another and is still working 20 years later. It got to work immediately on the observation of disks where exoplanets might be expected to form, and from there, the search for the black hole in the center of our galaxy, as we shall see in Chaps. 6 and 7.

It was while reflecting on the latter mysterious object that Eric Gendron came to equip NAOS with a unique capability. Eric was one of my students at the end of the 1990s. Drawn by his imagination, his experimental skill, and his considerate nature, I suggested he join our little No. 1 battalion using adaptive optics to combat blur, while No. 2 battalion was developing interferometry on the same front and for the same purpose—to use a military analogy! In 1995, he defended a fine doctoral thesis, whose ideas can be found today in many adaptive optics systems (Gendron 1995). Among his ideas, one is a wavefront analyser operating in the infrared. Now, 1 day or another, it would be important to take a closer look at the center of our Galaxy and its supposed black hole. In that case, there was no hope of implementing adaptive optics with a system which analysed the visible light from a star to inform the deformable mirror, since only infrared light can actually reach us from such regions. This argument finally convinced the ESO and Eric developed an infrared analyser called Rasoir in Meudon that would play a crucial role in the still distant victory of 2018 (Fig. 5.6).

Building the Interferometer

We would now have to move from the drawing board to the workshop and start building the many components of what could only be called "the complex interferometric system of the VLT". But the complexity that had so concerned the project manager Massimo Tarenghi now brought us all together. To cut costs, the initial project was slightly amended by dropping certain elements that were not considered essential.

The precision required in the construction of the interferometer system was formidable. At the Paranal site, the 8.2 m telescopes would be situated some hundred meters apart. The light from the star collected by the primary mirror

---

[22]The Cassegrain focus is located behind the primary mirror, which has a central hole to let the light through.

Fig. 5.6 The author with the team of young and enthusiast French astronomers, checking optics to fight against the blur, in the laboratory at Meudon (Paris Observatory) in 1995. *From left to right*: Jean-Marie Mariotti, François Lacombe, Pierre Léna, Olivier Marco, Daniel Rouan, Olivier Lai, Guy Perrin. Credit: Observatoire de Paris/LESIA

had to carried along a certain path to the point were the interference fringes would be formed, and the length of that path had to be known with an error less than a few billionths of a meter (nanometers). The only way to achieve this was to establish what engineers call an error budget. This is a table which distributes the final tolerated error over all the likely causes of error. Each of these, due to each of the many subsystems, must be analysed and reduced. To build this table, Bertrand and his team went ahead with dozens of calculations, and when necessary, laboratory measurements on whatever element was under investigation.

The various industrial contracts were eventually sent out, including those for the delay lines and the auxiliary telescopes. The latter stimulated lively competition between *Aérospatiale* in France and the Belgian company Amos in Liège. In the days when the Liège coal basin was in full production and supplied its iron and steel industry, Amos built locomotives and other machines to equip the mines. With the decline of mining activity from 1960, the company had to change course. The astronomical observatory in Liège was just getting involved in space missions to study ultraviolet radiation from stars. The satellite equipment had to be tested in vacuum, in large tanks simulating the hostile environment of space. After a major technical and managerial shakeup, stimulated by the astronomer Jean-Pierre Swings, the engineers at Amos changed their activity from locomotives to satellites. Twenty years later, they obtained the ESO contract to build the VLT's fragile auxiliary telescopes.

Alas! In July 1998, Jean-Marie Mariotti fell seriously ill and passed away just 3 weeks later (Léna 1998). There was a poignant ceremony in Garching, which I attended. Oskar, who had left the ESO to direct a large German solar research institute in Freiburg-en-Brisgau, was unable to hold back his tears. Jean-Marie would never see the inauguration of the VLT, on 5 March 1999, at the top of Paranal, and he would never visit the 'metro' tunnel that had just been completed. He wouldn't be able to contribute to the development of the European space project, the Darwin interferometer that was so important to him, and he would never know that it was abandoned in 2008. To perpetuate his memory, French astronomers would later set up the Jean-Marie Mariotti Center (JMMC) in Grenoble, which has become an international benchmark IT tool for processing fringe measurements.[23]

The ESO interferometer programme suddenly had no one to lead it, and the task was quickly passed on to a young, attentive to detail, generous, and hard-working German by the name of Andreas Glindemann. He remained in that position for the next 8 years, assisted by the young Belgian astronomer Françoise Delplancke, who would then take over from him. Bertrand, who had proven himself so brilliantly and contributed technically to saving interferometry, was called to other responsibilities within the VLT project.

The interferometer was scheduled to go into service at the beginning of the following decade, so the time had come to start working on the instruments which, supplied with the light from several telescopes, would analyse and measure the fringes. According to the rules established at the ESO, the task of designing and building these instruments was open to competing offers from European laboratories who would also have to provide the financing. The ESO would select the winners and sometimes ask several of them to work together. Then the work would be carried out in several stages: first on paper,

---

[23] The reader will find the beautiful poem written by Alisa Ridgway, Steve Ridgway's wife, at the site www.jmmc.fr/.

then building the instrument, and finally, testing it and setting it up on site in Paranal. Two instruments referred to as first generation—replaced in 2019 by something more up to date—were soon selected. The first, AMBER, would work in the near infrared, while the other, MIDI, was designed for longer infrared wavelengths, around 10 μm. In the latter range, one has access to cooler astronomical sources, such as the atmospheres of ageing stars.

To build these two instruments, French and German teams shared most of the work, but not without associating other European laboratories for certain tasks. All the know-how accumulated over more than a decade now could finally be put to good use. The interferometer community was at last mobilised to achieve a common goal that was full of promise and soon to become a reality. Everyone knew that there would be guaranteed nights of observation to reward these hard-working builders and their close colleagues. Naturally, this return on investment was a major motivation for the institutions ready to put up money, such as the CNRS in France, the Max Planck Society in Germany, and the *Istituto nazionale di astrofisica* in Italy, just as it was for the astronomers themselves.

With his team and in just 3 years, Andreas led the VLTI to its long hoped for fulfilment. During the night of 30 October 2001, two of the 8.2 m telescopes, Antu and Melipal, separated by a distance of 102 m, were coupled and obtained their first fringes, successfully detected and measured by a temporary instrument called Vinci.[24] Meanwhile, the two 10 m Keck giants in Hawaii had been successfully coupled and obtained their first fringes a few months earlier, on 12 March.[25] Apart from the blow to European self-esteem, subsequent scientific exploitation of the VLTI would in no way suffer from these few months' difference.

The first two egg-cup ATs were coupled with the same instrument Vinci in the night of 2 to 3 February 2005, thus delivering their first fringes.[26] The powerful instruments AMBER and MIDI became available 2 years later and their observations with the giants began over the following months and years. Of course, these required each to be equipped with adaptive optics, without which any information contained in the fringes would have been blurred. Four adaptive optics systems called Macao were installed by Norbert's team in 2003. These four Macao systems used the ideas developed in Hawaii for the

---

[24]The full interferometer system had been tested with success in February using the instrument Vinci, to which the light had been supplied by two small siderostats, not by the giant telescopes.

[25]www2.keck.hawaii.edu/1stLight/1stLight.html.

[26]VLTI First Fringes with Two Auxiliary Telescopes at Paranal, https://www.eso.org/public/news/eso0508/.

instrument PUEO by Roddier (2004). So at the output of each telescope, the seeing blur of each beam was corrected by a Macao, and then the light went on into the delay line tunnel to reach the instruments AMBER and MIDI. Although the latter were in full scientific use, they nevertheless had limited capacity. Whether the light came from the giants or the small telescopes, AMBER could only take light from three telescopes, and MIDI only from two. Neither AMBER nor MIDI could couple all four giants. It would be another five long years before we had an instrument capable of doing that.

At the observatory in Grenoble, they took up the challenge. At the ESO, it was suggested to build what was tentatively called a visitor instrument— a prototype which tests new principles without mutual commitment. The Precision Integrated-Optics Near-infrared Imaging ExpeRiment (PIONIER) was a superb instrument designed in just 1 year and set up in 2010 in the underground room where the four light beams were brought together. It received the beams, mixed them, and detected the fringes. It then measured the visibility and phase of each of the six sets of fringes, corresponding to the six possible combinations of the four giants (Le Bouquin et al. 2011. See also Zins et al. 2011). All these functions were carried out using an entirely new concept. The light was channelled in optical fibres. All the optical functions were carried out with tiny integrated optics chips, rather than using cumbersome mirrors.

The last step was taken on 17 March 2011. That night, all four giants Antu, Kueyen, Melipal, and Yepun were at last coupled together with the help of PIONIER. This was a tremendous achievement that had started out as a dream 30 years before! The way was now wide open over the coming decade to observe the immediate neighbourhood of the massive black hole located at the center of our Galaxy. We shall return to this in Chap. 7.

Parallel Paths

The reader will have noted the intense activity at the ESO between 1996 and 2005 to get the adaptive system NaCo up and running and prepare the coupling of the two first giants and then the two egg-cup auxiliary telescopes. Concomitantly, European institutes were building NaCo and the VLTI's crucial fringe detection instruments AMBER and MIDI. However, it was important to go on preparing for the future, an absolute rule in all research endeavour. Given the wealth of problems involved in getting an interferometer working optimally, several lines of investigation had been opened, in particular by the French teams in Nice, Grenoble, and Meudon, as well as in Germany and even in the ESO itself. All these investigations contributed at some point or another to the development of the VLTI. Inventive, fruitful, and sometimes rather curious, it will be worth our while discussing four of these before coming back to Paranal.

PRIMA and the Dual Field

There is an apparently insurmountable barrier for any Earth-based interferometer, due to the fact that the light which reaches it has had to cross an inhomogeneous and turbulent atmosphere. Adaptive optics provides a way round the seeing barrier, but it can do nothing against the rapid random displacements of the fringe system given by a pair of telescopes. On Mount Wilson, Michelson had been able to identify the bright fringes of the star Betelgeuse with the naked eye. But if he had taken a photo with an exposure of a few seconds, those fringes would have been irretrievably washed out. This displacement of the fringes is caused by the random variations in the light travel time, depending on the distinct paths it follows through the atmosphere to reach one or other of the telescopes. If the observed object is not bright enough to be able to detect and measure its fringes with a very short exposure, then it becomes impossible to measure the visibility (relating to the contrast) and the phase (the position of the fringe set) of the fringes. No hope, therefore, for an image, and no interferometric measurement of the details! But it is hard to accept such a limitation, making it impossible to observe objects that are too faint.

Now it so happened that, early on in our reflections in the 1980s, an idea came to the fore, initially suggested by Antoine and championed by Jacques Beckers during the detailed design of the interferometer. It was true that, even with all the light collected by the large telescope, the object that really interested us was too faint for the fringes to be measurable before being completely jammed! But quite by chance, there might just be a bright star in the immediate neighbourhood of this object. If the separation between the two objects were small enough, the light rays from each would follow almost exactly the same path through the atmosphere, undergoing the same perturbation and hence giving two sets of fringes displacing in the same way, at the same rate and with the same agitation. One of them, that of the bright star, would be easy to measure with a short exposure, while the other would remain invisible. The idea that interested us was to lock this second set of fringes onto the position of the first, the latter being permanently monitored. We could then do a long exposure on the second set of fringes, which was now stabilised and would not now be blurred by the long exposure. After a sufficiently long exposure, the fringe detector would have received enough light for the fringes to be measured. And that's all there is to it! With this idea, one could hope for a considerable improvement in sensitivity, by a factor of ten or even a hundred. Measurements of the visibility of the fringes and also their phase could then be used in aperture synthesis to build up an image of the faint object. The method has two names, each referring in some way to the underlying idea: 'dual field' or 'fringe tracking'.

Note here a certain resemblance between adaptive optics and interferometry. Thanks to an adaptive correction, a measurement made on a bright object B is able to remove the detrimental effect of the atmosphere on the light from an object F at a small angular separation from B but too faint. A long exposure on the latter then provides a deblurred image. It was Mike Shao and Mark Colavita who first grasped the power of this method when applied to interferometry, using it for their small astrometric interferometer in the 1980s (Mozurkewich et al. 1988. See also Shao and Colavita 1992). Calculation shows that the distance between the star and the faint object should not exceed five to ten seconds of arc, depending on the state of the atmosphere. Hence, with his working group at the ESO, Jacques Beckers drew up the project with an interferometer field of eight seconds of arc, and this determined the diameter of the mirrors in the 'metro'. The financial restrictions following the crisis reduced this field to two seconds of arc, and by an incredible stroke of luck, this angle of two tiny seconds of arc was enough to find a bright enough star and come really close to the black hole, when observing the center of the Galaxy.

This was the icing on the dual field cake, if I may so express myself! Here were two celestial objects B (bright) and F (faint), very close together, in fact separated by less than two seconds of arc, producing a set of fringes on the detector. The shift between these two sets depends only on the length of the baseline and the angle separating B and F—these two quantities alone fix the difference in travel time for the light from B and the light from F. To use the jargon, this is the relative phase shift of the two sets of fringes. If the exposure is long enough for the fringes of the faint object F to become clear, then this phase shift can be very accurately measured. We call this a differential measurement. Since the baseline is very accurately known, the angle between B and F is thus determined. Thanks to the dual field, the interferometer can achieve a level of astrometric accuracy well beyond its resolution limit.

By the time the crisis finally blew over in 1996, everyone had realised that the exoplanets that were just being discovered would be a major topic of investigation by the future inteferometer. A young German, Andreas Quirrenbach, put forward a project that would make use of this very accurate astrometry. This project, called Phase-Referenced Imaging and Micro-arcsecond Astrometry, or PRIMA, was adopted by the ESO. The studies and first realisations were carried out under Andreas Quirrenbach and pursued by Françoise Delplancke (van Belle 2008). In Chaps. 6 and 7, we will see how both exoplanets and black holes would benefit from an astrometric accuracy almost three hundred times better than the resolution of the interferometer.

From Fringes to Image?

Obtaining and measuring the visibility of beautiful interference fringes, and what's more for several baselines, is of course a great start. Being able to accurately determine the relative phases for each baseline is even better. And indeed, this determination is a necessary condition for being able to reconstruct a genuine image of the target object on the basis of these sets of fringes. When I discussed this subtle point, called aperture synthesis, in Chap. 4, I mentioned that in 1987, when it was decided to go ahead with the VLTI, none of the many small optical interferometers then operating had yet managed to do this. However, we were still sure the VLTI would be able to do this because we had the example from radioastronomy,[27] extended by Gerd's work on speckle. For the radioastronomers, aperture synthesis held no mysteries, no difficulties of principle, as demonstrated daily by the beautiful images obtained by the Very Large Array (VLA). We also knew that, whatever the wavelength, we were still dealing with light and hence the same laws of physics, translating into similar equations and similar solutions.

In 1996, the situation changed somewhat. This appeared in the New York Times:[28]

> In February, Dr. John E. Baldwin and his colleagues at the university's Mullard Radio Astronomy Observatory in England published the first detailed pictures ever obtained by any optical telescope of the double star Capella, which lies 40 light-years away. (A light-year is the distance light travels in 1 year, at the rate of 180,000 miles per second.) The two stars in the Capella pair are only a little more than one million miles apart, far too close to be seen from Earth as separate objects by any conventional optical telescope, including the huge Keck Telescope in Hawaii, the world's most powerful, and the Hubble Space Telescope. The remarkable resolution, or sharpness, of the Cambridge group's images of Capella was achieved using three optically linked telescopes of only

---

[27]The general name given to this elegant method is phase closure. It would no doubt take us a little too long to try to explain it here.

[28]New York Times, 30 April 1996.

modest size at a site in England that would have been spurned by builders of large conventional telescopes. Cambridge is barely above sea level and is subject to England's notoriously unstable weather.

And so it was that the decisive step was taken experimentally in this prestigious university where Isaac Newton had taught and Thomas Young had studied, where Martin Ryle had founded aperture synthesis, where pulsars were discovered, and where Martin Rees was a professor. John Baldwin, a few years my senior, was one of the heirs of Martin Ryle, the Nobel prizewinner that we met at the Cavendish Laboratory in Cambridge. Although his country was not a member of the ESO, joining only in 2002, we brought John into our pre-liminary reflections on the matter. This phlegmatic and rigorous Englishman, respected scientist, member of the Royal Society, had taken 10 years to build his interferometer, so aptly called the Cambridge Optical Aperture Synthetic Telescope (COAST). It included five small telescopes, rather than two. It could thus simultaneously observe with several interferometric baselines,[29] a necessary condition to be able to produce a 'true' image using aperture synthesis despite the random phase changes imposed on the fringes by the atmosphere. The subject studied by John and his small group was Capella, a bright double star in the constellation of Auriga, the Charioteer, visible in our winter sky.

A further equally historical image of this double star system was obtained in the near infrared on 25 October 1997. The two components of Capella are clearly visible, precisely located with an angular separation of 55 milliseconds of arc,[30] which is about one twentieth of the size of the seeing disk. Recall that, in 1920, Albert Michelson published the first interferometric observation of an object located outside the Solar System using his interferometer observations with the Mount Wilson telescope. And the subject was this same Capella binary star, although his method could not of course provide an image in the strict sense. He simply measured the angular separation of 55 milliseconds of arc between the two stars. A month later, with four telescopes, John's group obtained the first image of the surface of a red giant, the bright star Betelgeuse in the constellation of Orion. This was also the first star whose disk Michelson had resolved! And this time, the star Betelgeuse was no longer viewed as a geometrical point by the Earth-bound observer. Going a step further than

---

[29]Five telescopes A, B, C, D, E yield $(5 \times 4)/2 = 10$ combinations, each providing another baseline: AB, AC, AD, AE, BC, BD, BE, CD, CE, and DE. The fifth telescope was not set up until 1998.

[30]www.mrao.cam.ac.uk/outreach/radio-telescopes/coast/coast-astronomical-results/.

Michelson's measurement, John's image showed a circular disk with some details.

John Baldwin's achievements came at just the right time to give support to the options selected for the VLTI. We just had to adapt the method to telescopes with a diameter almost a hundred times greater. And this was achieved in 2008, using the three egg-cup ATs on the one hand and the three giants on the other to form different interferometer baselines for the instrument AMBER in Paranal. The result was a genuine image of a double star system, the binary star known as Theta$_1$ Orionis C, located in the beautiful constellation of Orion. And so it was that the team working with Gerd, which had been working on this star for 20 years using his image reconstruction technique, could finally announce that aperture synthesis was now possible in Paranal (Kraus et al. 2009). In 2019, it is common practice to obtain true images by aperture synthesis, as I shall describe further when discussing exoplanets and the galactic black hole.

Optical Fibres, Pipes for Light

When the light from a star arrives in the underground room at Paranal and enters the instrument that measures the fringes, it has already undergone about twenty reflexions on mirrors. The first of these occurs on the large primary mirror of the telescope being used, and it is followed by further reflexions which send the light to its underground path and the mirrors of the 'metro' and a few others, before it finally enters the instrument room. However carefully these many mirrors are polished and made as reflective as possible, kept safe from the desert dusts and cleaned, such a large number of reflexions inevitably attenuates the light beam. A large part of the light energy—sometimes more than 90% in the case of the VLTI—is inevitably lost. It goes without saying that everything possible must be done to reduce the loss of these precious grains of light.

Now, at the beginning of the 1980s, telecommunications were transformed by a technical revolution that changed the world. Until then, telephone communications between two continents had used radio waves which passed via costly geostationary satellites. The latter were launched into positions at an altitude of 36,000 km above the Earth's surface, and served to relay the waves carrying the messages from one point on the surface to another. But then came lasers and optical fibres, tiny transparent pipes made of extruded silica glass covered by a protective sheath and laid across the ocean floor to connect the continents. Simplifying slightly, the basic idea is as follows. At one end of the fibre, a laser emits a beam of light that has been modulated in some way by the message to be transmitted. At the other end, the light is received and detected, then the message is decoded and sent to the receiving telephone. These fibres are easy to produce and they have literally transformed intercontinental telecommunications, including television communications, drastically reducing their cost. Everyone can see the results today and benefit from them.

So in order to transport the light in our interferometer, would it not be possible to replace the sophisticated and costly sets of mirrors by just such a magic pipe? This was the question I asked myself when we were discussing the best kinds of interferometer layout at the ESO in the 1980s. After all, hadn't radioastronomers long been transporting waves along hollow metal pipes, which are easy to make for the wavelengths they work with? I had thus suggested that one of my young students, Vincent Coudé du Foresto, could work on this idea as a subject for his doctoral thesis. At the time, in the pleasant town of Rennes in Brittany, there was an inventive chemist and professor at the university, Jacques Lucas, and the enterprising founder of the company *Le Verre fluoré*, Gwendaël Mazé, who was also a genial advocate of the local language, Breton. Uniting their talents, they had developed special glasses and were producing optical fibres capable of transmitting infrared light. Better still, they were able to make what are known as single-mode fibres, the most suitable for long-distance communications. When light propagates along a fibre, it can follow many different paths, reflecting off the walls of the fibre more or less often depending on the path. These multiple reflexions lead to a very small loss of energy, but more importantly to a loss of the exact phase of the light wave. The phase of the wave is well defined when it enters the fibre, but no longer when it leaves. It would be a disaster if such a multi-mode fibre were used in an interferometer, where the phase measurement is crucial! But a single-mode fibre has a truly tiny diameter, of the order of the wavelength of the light that it must carry. Finer than a hair, it can carry the light wave without losing the phase between input and output.

Hence, by carefully examining the use of Gwendaël Mazé's single-mode fibres in the basement of the ESO building in Garching, the young and enthusiastic Vincent bravely set out to find a fibre coupling device that could replace the sets of mirrors in an interferometer. And as often happens with young researchers with alert minds not yet cluttered by too much knowledge— or indeed preconceptions—like those of their seniors, and whose energy appears to be without limit, Vincent actually succeeded in doing two things at once. Not surprisingly, he showed experimentally that fringes could be obtained by transporting light in this way. But he also made an important discovery, proving that the degradation of the fringes due to atmospheric turbulence was significantly reduced by the ability of the fibres to filter this turbulence.

Jean-Marie Mariotti and Stephen Ridgway, my fellow travellers for over a decade, helped him to clinch the demonstration by allowing him to set up his optical fibres at the Kitt Peak Observatory in Arizona, where Steve was working. Ten years earlier, when designing his own interferometer, the physicist Charlie Townes from Berkeley in California had used the configuration of the MacMath Solar Telescope, built for observation of the Sun. He had combined the light collected by two medium-sized mirrors (0.8 m), 5.5 m apart.[31] In Vincent's case, the focal point of each mirror was coupled with a fibre, the two fibres were brought to the same infrared detector, and the fringes resulting from the mixture of the two beams were then measured. The telescopes were aimed at bright stars and the experiment, entitled Fiber Linked Unit for Optical Recombination (FLUOR), was a great success, perfectly demonstrating the principle of fibre recombination that he had established.

After this brilliant demonstration, one beautiful summer's day in 1994, Vincent presented his doctorate at the University of Paris VII. At this point, Vincent's work was taken over by Guy Perrin, another doctoral student who had joined our group. Making use of our connections with colleagues in the United States, Guy set up the FLUOR instrument at the focal point of the little interferometer IOTA, built in Harvard by Wesley Traub, who would later lead NASA's exoplanet exploration group, then based on Mount Hopkins in Arizona, among the cacti and the ocotillos, in a landscape that would have been perfect for filming a western.[32] With this new instrument, Guy refined Vincent's demonstration. He defended his doctorate at the end of 1996, just as the ESO resolved the crisis that almost put an end to our war on blur. With renewed optimism, Guy and his friend Sacha Loiseau published a nice paper about the prospects for optical interferometry, with the subtitle *Ombres et lumières dans l'univers* (Light and shade in the Universe), in the magazine *La Recherche* (Loiseau and Perrin 1996). But more importantly, he published a series of observations made with IOTA in the infrared using interferometer baselines of 15–20 m. He determined the diameters of giant stars which were of the order of 10 to 20 milliseconds of arc. The remarkable accuracy, to within

---

[31]Later called the McMath–Pierce Telescope, in honour of two American astronomers, this magnificent telescope is still in service in 2019. The two mirrors mentioned here are flat coelestats, permanently orienting the Sun's beams in a fixed direction, toward two concave mirrors forming the image. This is equivalent to two telescopes pointing in parallel at the same object.

[32]www.researchgate.net/publication/2667269_The_FluorIota_Fiber_Stellar_Interferometer.

a millisecond of arc, confirmed once and for all the advantage in using single-mode optical fibres to transport and recombine the light beams (Perrin et al. 1998).

Guy and Vincent played a key role in the rebirth of the VLT interferometer after 1996, but unfortunately without Sacha, who left us to found his own company.[33] Vincent became one of the outstanding figures of French interferometry. A professional astronomer who never forgot the importance of amateur astronomers, an excellent lecturer, a keen observer of total eclipses of the Sun, he certainly knew how to share his enthusiasm with all kinds of audience, both in France and in Africa. In 1996, once the crisis had passed, the ESO turned its attention to the very first instrument that would measure the first fringes from the VLT, as soon as the giants were coupled together. The instrument it was financing was of critical importance for the future of interferometry. Working in our team at the Paris Observatory with Guy, Vincent suggested that the ESO should implement the single-mode fibres he had shown to be so efficient. They could be used with the instrument he had called Vinci as a sign of admiration for the great man, an admiration shared by his partner Saskia when they called their son Léonard.

Vinci was adopted by the ESO, so we had to help Vincent build it and set it up as quickly as possible in Paranal. Among the students attending my masters course, I had noticed one in particular, a rather shy engineer whose face lit up whenever I was talking about interferometry. When the time came for Pierre Kervellat to choose a research subject, as first research experience and then in preparation for a doctorate, I didn't hesitate to suggest that he get involved with work on Vinci for the VLTI. As we had hoped, in Paranal, at the end of 2001, the instrument Vinci immediately proved that it could increase the accuracy of our interferometric measurements by almost a factor of ten, just by the way it filtered atmospheric turbulence. Pierre's work had a considerable impact, because he used his interferometric measurements to improve the accuracy of distance measurements in the Universe based on the pulsations of the so-called Cepheid variable stars (Kervella 2007). Then he left for Chile where, for the next 2 years, he spent 80 nights a year forming fringes with Vinci for astronomers using the combined light of two of the giants. Meanwhile, Vincent turned his attention to those other worlds, the exoplanets, and we shall meet him again shortly.

---

[33]Mauna Kea Technologies was founded by Sacha Loiseau, who chose this name in order to remember his observation missions in Hawaii. In 2019, MKT has become a highly innovative company in the field of medical optical equipment. The technical director is François Lacombe, also from our anti-blur team. Who knows where astronomy may take you! www.maunakeatech.com.

Guy, whom we encountered at the very beginning of this story, changed course and left us after his doctorate. This brilliant graduate of the *École polytechnique* was recruited by the Ministry of Defence to supervise certain classified projects. And when this 3-year contract came to an end, he hesitated to come back to astronomy, given the lucrative offers from industry that could never be matched by a university research post. However, it was astronomy that won the day and the reader may easily imagine how pleased I was to take him back into our team, especially now that the VLT interferometer was back on the rails, if I may put it like that! He would often tell me that he never regretted this choice, result of a long reflection. The conclusion of his doctoral thesis suggested as much:

> The recent signature of an agreement committing the CNRS to the construction of the future large European interferometer, the VLTI, may perhaps invite the construction of a larger scale and more modern fibre recombination unit. Optical interferometry may then at last enter an era of astrophysical discoveries on a par with what has been witnessed in radio interferometry.

And indeed, 20 years later, Guy would be among the first to put his name to the discoveries made regarding the black hole at the galactic center!

After Vincent and Guy, this line of work with optical fibres continued with the development of PIONIER, which I have already described. PIONEER was the successor to Vinci, made in Grenoble by a team including the young and brilliant Jean-Philippe Berger, who would 1 day be in charge of the interferometer in Paranal. PIONIER was an extraordinary instrument, and the team in Grenoble brought together a number of very talented people. I have already mentioned Pierre Kern, one of the pioneers of adaptive optics, who founded a company to build it, and Fabien Malbet, the man who worked on protoplanetary disks, also passionate about educational issues. And there were a whole host of other contributors to PIONIER that quite unjustly I am unable to name, even though I knew many of them, including some women, although women are sadly lacking in this story. So let's not forget Myriam Benisty, a specialist on protoplanetary disks, and Karine Perraut, expert on instrumentation. All this talent went into the construction of PIONIER, an instrument made entirely from integrated optics and not much bigger than a matchbox, inside which light from the depths of the Galaxy is propagated, split, recombined, modulated, and polarised, then made to interfere before ending its journey on detectors transforming the infrared fringes into electric charges and currents.

This spelt the end, at least in part, for bulky optical benches and mirrors with their unfortunate tendency to gather dust! From now on, these ingenious light-carrying pipes, the single-mode fibres, would become the norm at the focal points of optical interferometers.

OHANA

Let us say one more word on the twists and turns of those heady years! During the crisis at the ESO in 1993, we were in such a state of despair to see the VLTI delayed or even dropped completely that we set about exploring other options, just in case. Now at the top of the extinct volcano of Mauna Kea, on the Big Island of Hawaii, several very large telescopes were either already operating or under preparation. These included the 3.6 m Canada–France–Hawaii Telescope (CFHT), the two 10 m giants of the Keck Telescope, and the 8.1 m Gemini Telescope of the National Science Foundation which was about to be inaugurated in 1999. A Japanese 8.2 m telescope called Subaru was also decided in 1991 and would be located on the same summit. A rather mad possibility came into view: the idea of connecting two of these giants by a tiny optical fibre, carrying the light from the stars, barely buried beneath the volcanic soil, to avoid any environmental impact on the mountaintop, to a common focal point where fringes could be formed (Mariotti et al. 1998). To my great surprise, my colleagues directing these observatories, in particular the Frenchman Gérard Lelièvre at the CFHT and the francophile and accomplished pianist Fred Chaffee at the Keck, were keen to support this mad idea. They would do everything in their power to ensure that the cooperation agreement we had signed for this purpose would be implemented without delay.

Several members of the cast came from Meudon, where Guy and Vincent, always first on the starting blocks whenever the word 'fibre' was mentioned, immediately took up the challenge. Olivier Lai became involved in adaptive optics in 1990. He loved the life in Hawaii and stood out at the CFHT for his energy and curiosity. Julien Woillez, who had just graduated from the *École polytechnique* and was afraid of nothing, also joined up. The instrument would be called OHANA, an acronym for Optical Hawaian Array for Nanoradian Astronomy.[34] The word *'ohana* also means 'family' in Hawaiian, so it would hardly have been possible to find a better name for this French–Hawaiian

---

[34] As we saw earlier, the radian is a unit for specifying angles, equal to about 60°. In interferometry, the angles to be investigated and resolved are very small, and indeed the prefix 'nano' means 'billionth'.

Fig. 5.7   OHANA project (2000). *Left*: At 4200 m altitude on the summit of Mauna Kea (Hawaii), Guy Perrin and Olivier Lai plan to set up the optical fibers which will connect the Canada–France–Hawaii telescope (dome at left) to the Gemini 8-m telescope (middle). *Right*: The drums with 2 × 300 m of fluoride glass optical fibers (Le Verre Fluoré, France), which will carry the light from each telescope to mix it and create the fringes. Credit: Pierre Léna and G. Perrin

project, which sought to bring together several huge telescopes (see Fig. 5.7). We only partially achieved our goal several years later (Perrin et al. 2006), but it was good training for the team in the art of interferometry. Guy in particular became an expert on the accuracy of interferometric measurements using optical fibres. In the past, a given measurement could estimate the fringe contrast to within 10%, while now it could estimate to within a thousandth! (Perrin et al. 2004). This work was a good preparation for the very difficult exploration of the galactic center that would be carried out with the VLTI. Meanwhile, Julien, who had earned his interferometry qualifications the hard way, digging in the volcanic ash at the top of an extinct volcano in Hawaii to lay his optical fibres, would end up in Garching in 2018, in charge of the GRAVITY instrument.

In the 2030s, nothing would stop us from reviving the idea of OHANA in Hawaii, once the pressure on the use of the large telescopes in Hawaii falls off with the increasing availability of even larger telescopes, thereby extending to the northern hemisphere the possibilities of interferometric observation with giant telescopes that are carried out by the VLTI in the southern hemisphere. I do believe that, down at the observatory in Nice where he is now working, Olivier Lai still dreams of doing this.

But after the many adventures I have just recalled, let us not get too far ahead of ourselves. Let us get back to essentials. In the year 2000, the adaptive optics system NaCo was set up on Yepun at Paranal. Then in 2001, Antu and

Melipal were coupled to form an interferometer, freed from the seeing barrier by the adaptive optics system Macao. We had just arrived in the twenty-first century. Thirty years had gone by since the beginning of our war on blur, and now our two champions in this endeavour, so complementary to one another, had been developed and launched on their careers. In their company and over the next two chapters, I will take the reader far out into the Galaxy to visit these other worlds and our very own massive black hole.

6

Images of Exoplanets

There are more things in heaven and Earth, Horatio,
Than are dreamt of in your philosophy—William Shakespeare, *Hamlet*

Are there other worlds in the Universe, outside our own Solar System, that look in any way like our own dear Earth? Six centuries before the common era, the Greek thinker Democritus had already formulated this question, and many others have done so since then, as mentioned at the beginning of this story. This purely speculative question has followed us down through the centuries. Advocates and opponents have wrangled over philosophical arguments, and in the West also theological ones, in the days before science came into its own.[1] In contrast to the Dominican Thomas Aquinas, the bishop of Paris, Etienne Tempier (ca. 1210–1279), maintained that the divine omnipotence had no reason to limit itself to the creation of a single Earth, and that it was not up to humans to decide one way or the other. The debate went on, religious or otherwise, but it was not until the twentieth century that our means of observation could begin to tackle the question in a less abstract way, other than through pure speculation. Then, at the very end of the last century, astronomers began to search in earnest, the first exoplanets were located, and the war on blur played a major role. An extensive new field of exploration had opened up. Perhaps it would prepare us for a new way of looking at our own little planet, and indeed at humanity itself?

---

[1]Arnould (2017). In his book, the scientist and theologian Jacques Arnould offers a detailed history and a 'thought experiment' which he formulates in modern terms.

© Springer Nature Switzerland AG 2020
P. Léna, *Astronomy's Quest for Sharp Images*, Astronomers' Universe,
https://doi.org/10.1007/978-3-030-55811-6_6

A Fruitless Quest

Just the idea of photographing a planet directly in its orbit around a distant star, even one as big as Jupiter, used to seem like a challenge that could never possibly be met. But it was hoped that one might at least detect some indirect consequence of its presence, by observing the motion of its star. The latter would be bright enough, and like all the other stars in the Galaxy, it would be in motion relative to the others. Subject to the weak gravity of any planet that might happen to be revolving around it, the star would zigzag slightly across the sky rather than following an absolutely straight trajectory. This zigzag would serve as a kind of observable 'signature' of the presence of a planet. Assuming, for instance, that the planet had a similar mass to Jupiter, one can calculate the amplitude of the zigzag and decide whether or not it would actually be observable.

A candidate for such a measurement turned up in 1916. Barnard's star, named after the astronomer E.E. Barnard in Chicago, is a mere six light-years from the Sun. Situated just a little further than the nearest star Proxima Centauri, its apparent motion across the sky is very fast. On a photographic plate, its position can be seen to change a lot from 1 year to the next, in fact ten seconds of arc every year, relative to the most distant stars on the photo, which appear motionless due to their great distance. The Dutch astronomer, Peter van de Kamp, who had been observing Barnard's star at regular intervals since 1938, claimed to have detected a zigzag in its motion. In his view, this proved that there had to be a planet of similar mass to our own Jupiter. As a young researcher in the mid-1960s, I well remember the lively debate during an astronomy conference I attended, with a heated discussion about the validity of van de Kamp's observations. Had astronomers at last found one of those 'other worlds' that Democritus had imagined? More careful and repeated observations 10 years later refuted the existence of any large planet in orbit around this star. But just the idea of a planetary system around a neighbouring star inspired a rather crazy British project called Daedalus, between 1973 and 1978, that was supposed to actually send an exploratory probe with nuclear propulsion right out to Barnard's star, on a journey that would take 50 years. Thirty-five years later, a new project came up in a somewhat different context. This project, called StarShot, is aimed at Proxima Centauri, our nearest stellar neighbour, now known to have at least one exoplanet.[2]

---

[2] en.wikipedia.org/wiki/Project_Daedalus. See also the StarShot project (2016): https://en.wikipedia.org/wiki/Breakthrough_Starshot.

A First Surprise

By the middle of the twentieth century, we had a better understanding of the conditions under which stars form, and the hypothesis of protoplanetary disks was gaining evidential support. I have already stressed that this study featured among the main objectives assigned to Europe's Very Large Telescope when the decision was finally made to build it in 1987. Because it was so close to the Earth, the star Beta Pictoris was the obvious target. The orbiting infrared telescope IRAS had identified an excess of infrared radiation from the vicinity of this star. Earth-based observations in 1984 showed that there was a disk around the star and that this was the source of the radiation. During the 1990s, the Hubble telescope provided images in the visible part of the spectrum,[3] with a resolution of 0.1 seconds of arc—this is the diffraction limit of this 2.4 m spaceborne telescope. Some excellent infrared images obtained with Earth-based telescopes equipped with adaptive optics, as described in Chap. 5, showed both this scattering and the emission by the cold dusts of the disk. The question was, did the disk around Beta Pictoris, now considered exemplary, actually show any signs of planet formation? During the 1990s, the question raised by Democritus, which focused attention on other worlds and the living creatures that might inhabit them, became a hot topic in the astronomical community. For its part, NASA resumed its exploration of the planet Mars with the ambition of detecting signs of life. In 1989, the astronomer David Latham at the Harvard College Observatory (Massachussets) published an observation of the brown dwarf HD114762, a very low mass star which appeared to have a companion that might be another brown dwarf or a giant planet. Things were getting interesting, although the fact that it was an exoplanet was only confirmed in 2012.

An Unexpected Discovery

In 1991, something totally unexpected happened. A young Polish radioastronomer working in the United States had specialised in the study of pulsars, extremely dense objects made up almost entirely of neutrons. These were the remants left over after the explosion of a star, and the emission of high energy particles, characteristic of a pulsar, was first postulated; by my Italian friend Franco Pacini, who had been so decisive in getting Italy to join the ESO. Quite

---

[3] Heap (2017), cited by Kalas, VLT Opening Symposium, European Southern Observatory, 1999.

by chance, using the giant Arecibo radiotelescope in Puerto Rico, Aleksander Wolszczan discovered two planets in orbit around the pulsar PSR B1257+12, situated some 2300 light-years from Earth. He determined their masses to be slightly greater than the mass of the Earth and their orbital periods to be a few months. At this distance, given the rather low masses of the objects, they were not discovered in the same way as those mentioned up to now. Such methods would never have been able to detect them at this distance. So how was it possible?

In fact, a pulsar is like an exceptionally accurate clock. Its ticking can be very precisely measured by the radio waves it sends to Earth. And it is extremely sensitive, through gravitational effects, to the orbital motion of one or more planets that may be revolving around it, even if they are not very massive. Presented in half a page at a conference organised by the American Astronomical Society with a title that ended with a suitably cautious question mark (Wolszczan 1991) and published shortly afterwards, the news went round the world. These were the first exoplanets of terrestrial mass to be discovered, where they were not at all expected, and by a method that was surprising to say the least! And it is worth mentioning that, despite a systematic exploration over more than 20 years, only four other pulsars have been found to have planets going around them. Note also that it is no easy matter to explain the presence of planets after the explosion of the star at the moment when the pulsar is formed. Could it be that these objects were somehow torn from a 'conventional' planetary system and were just wandering through space before simply being captured by the gravitational field of the pulsar? This is why the discovery of 51 Peg b in 1995 is considered to be the first discovery of a 'genuine' exoplanet.

Space Returns to the Forefront

The astronomical community soon swung into action. On 18 March 1992, a forum on extrasolar planets was held in Meudon.[4] Some of the speakers were convinced of the future importance of the subject and had already got involved. Among them, Jean-Marie Mariotti was investigating the potential contribution of interferometry. Jean Schneider was another enthusiast, with a keen interest in and remarkable knowledge of philosophy. He identified fourteen possible ways to look for such planets. Looking even further ahead with his colleague Alain Léger, he also started thinking about the possibilities

---

[4]Author's own archives.

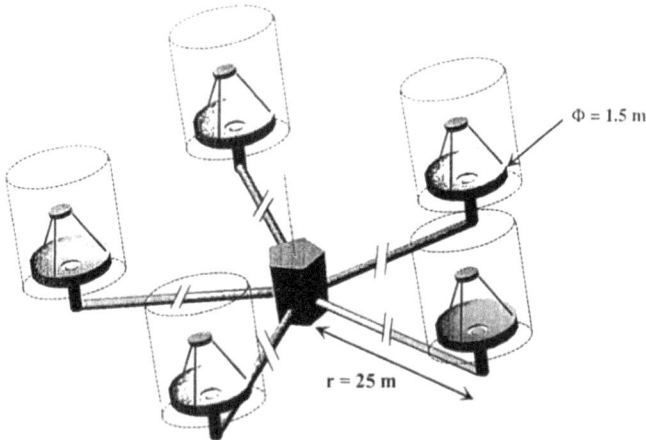

Fig. 6.1 In 1996, 1 year before the discovery of the first exoplanet, Alain Léger, Jean-Marie Mariotti, and others proposed the realization of the space interferometer DARWIN, to work in the infrared. Five 1.5-m telescopes were to be placed in space, sending light and forming fringes at their common focus, located at the center of the structure. Here, no long delay lines are necessary. Source: Léger et al. (1996)

for detecting life. In the autumn, preparing its long term programme Horizon 2000+, the European Space Agency organised a meeting called *Targets for space-based interferometry*.[5] Bernie Burke, a physicist and astronomer from Boston, presented NASA's reflections on this issue over the previous decade. The title of his talk included the term 'exoplanet' now in common use, doubtless for the first time. Once again, Jean-Marie spoke for the advantages of existing interferometers and the future VLTI, while Antoine put forward his projects for the Moon. And why shouldn't Europe envisage an interferometer in space?

A project began to take shape,[6] taking into account the goals set out by NASA in 1992 in its programme entitled *Towards Other Planetary Systems*. An interferometer called Darwin, operating in the infrared so that it could detect ozone in a planet's atmosphere,[7] was proposed in 1995 (Vakili and Loiseau 1995) and again in 1996 (Léger et al. 1996) for inclusion in the programme of

---

[5] Targets for space-based interferometry, ESA SP-354, December 1992.

[6] *La détection des planètes extrasolaires*, L'Astronomie, July 1993. Articles by J. Schneider, A. Vidal-Madjar, and J.-M. Mariotti.

[7] The ozone in the Earth's atmosphere (i.e., the molecule O_3) is produced by dissociation of dioxygen molecules O_2 by solar ultraviolet radiation, followed by a recombination involving three oxygen atoms O. If there were no life on Earth, the dioxygen would quickly disappear, and so also would the ozone. The latter is thus considered as a likely marker for the existence of a biosphere in which photosynthesis is occurring.

the European Space Agency (see Fig. 6.1). It would include five 1.5 m telescopes arranged in a petal formation with baselines of a few tens of meters. Jean-Marie and Alain Léger put in a huge effort, attracting brilliant students like Bertrand Mennesson and Marc Ollivier. Obviously, the name was chosen in homage to Charles Darwin for his theory of evolution, a theory which may well turn out to describe a universal law. The name thus indicated the objective: to observe exoplanets and search for signs of life.

Unfortunately, this magnificent mission was ahead of its time. The stream of discoveries of exoplanets after 1995 completely transformed the landscape. In addition, Darwin was technologically very difficult and hence costly. The Darwin mission was dropped in 2009, just as NASA dropped its project Terrestrial Planet Finder in 2011, the last manifestation of the Space Interferometry Mission (SIM) designed earlier by Mike Shao at the Jet Propulsion Laboratory in Pasadena.

A Second Surprise and Its Consequences

At the beginning of this tale, at the top of Paranal with Anne, I mentioned a very important event on 6 October 1995, namely, the discovery of the first extrasolar planet orbiting around an ordinary star, a discovery made by two astronomers in Geneva, Michel Mayor and the young Didier Queloz.[8] The planet cannot be seen directly in an image. Only its gravitational influence on the motion of the star was actually measured, by observing the regular, periodic shifts of the spectral lines emitted by the star's atmosphere. This indirect detection method is called velocimetry, since it measures variations in velocity. The star 51 Pegasi is situated 51 light-years from Earth and it is just bright enough to be seen with the naked eye. Its planet, since named Dimidium, is a gaseous giant planet with an orbit very close to the star, to everyone's surprise, twenty times closer than the Earth to the Sun. Its 'year' only lasts four and a half of our own days. It has a mass equal to about half of Jupiter's mass (hence the name), which would be about 150 times the mass of the Earth.[9] This discovery will lead to a Nobel prize in physics, given to these two astronomers in 2019.

---

[8] In 1989, it was announced that an object with a similar mass to Jupiter had been found in orbit around the star Gamma Cephei. The observation and its interpretation were contested, then finally confirmed in 2003 (Hatzes et al. 2003).

[9] The full story of this discovery, its initial contestation, and the final triumphant confirmation are very well told in Mayor and Frei (2001).

While the discovery of a planet in orbit around a pulsar in 1992 had little consequence, the identification of 51 Peg in 1995 mobilised the astronomical community in a quite extraordinary way. A quarter of a century later, just as I come to finishing this book, a total of 4001 exoplanets have been detected, confirmed, and catalogued.[10] They are all located in our own galaxy, usually no further than a few hundred light-years from Earth. Many orbit around stars that can be seen with the naked eye or a good pair of binoculars.

This would not be the place to describe the wide range of methods that have been used in these explorations (Lagrange 2019), most of which were already under discussion in 1992, as Jean Schneider explained at our forum in Meudon. And neither could I discuss in any detail the huge body of knowledge that has been derived from these discoveries, or the questions they have raised. I shall merely skate over the surface. So, continuing with the theme of this book, I shall focus on the contribution to this knowledge made by our victory over the seeing limit, as achieved today at Paranal. The reader may consult the tables A and B at the end of the book for certain quantitative details regarding the perception of distances and angles on the sky, and also the anti-blur capabilities of the telescopes that have been used in successive investigations of exoplanets.

Images of Exoplanets at Last!

As soon as the first light was collected by the NaCo adaptive optics instrument at the end of 2001 (Brandner et al. 2002), Anne's team and her many collaborators spent long nights at Paranal. The aim was not just to obtain images of the stellar disks. Now that the blur had been removed, the hope was to make out an exoplanet directly on an image whose sharpness was only limited by the diffraction limit of an 8.2 m telescope, that is, something like forty milliseconds of arc. The distance between the Sun and the planet Jupiter is 5 astronomical units, that is, five times the distance between the Sun and the Earth. If the observed object was at a distance of a hundred light-years from the Earth, this separation of 5 AU would subtend an angle of 160 milliseconds of arc, which would be easily resolved by the telescope Yepun. Observing the several hundred bright enough stars closer to the Sun than a hundred light-years, it was therefore reasonable to expect to find a planet which, in an image, would appear distinctly separate from its star. Naturally, apart from having a sufficiently sharp image, a coronagraph would also have to be used to mask

---

[10]See the complete and up-to-date site managed by Jean Schneider at the Paris Observatory: https://exoplanet.eu/diagrams/.

the light from the star, otherwise the brightness of the star would completely drown out such a tiny source of light. Indeed, the brightness of a star could be millions of times greater than the faint light scattered by a nearby planet.

The quest was rewarded in 2004. The star 2M1207 is known rather unappealingly as a 'brown dwarf'. These are cold and peculiar stars with a very low mass, but somewhat more than an exoplanet, radiating mainly in the infrared. It was thus necessary to use the infrared wavefront analyser, the one devised and built by Eric Gendron in the form of his prototype Rasoir and now installed on NaCo. The image revealed the presence of an exoplanet situated at 0.7 seconds of arc from the star, and with an estimated mass of about 5 times the mass of Jupiter (see Fig. 6.2). This was a world first, obtained by Anne and her young student Gaël Chauvin, published under the cautious title *A giant planet candidate near a young brown dwarf*. It remained to show that the proximity of the two objects in the image was not just a chance association, in other words, that they were indeed revolving around one another in their shared gravitational field (Chauvin et al. 2004). The result was finally confirmed a year later(Chauvin et al. 2005). So at last, adaptive imaging had shown that an exoplanet could be seen directly. Some critics, nitpicking or perhaps just jealous, argued that this brown dwarf was not really a 'true' star,

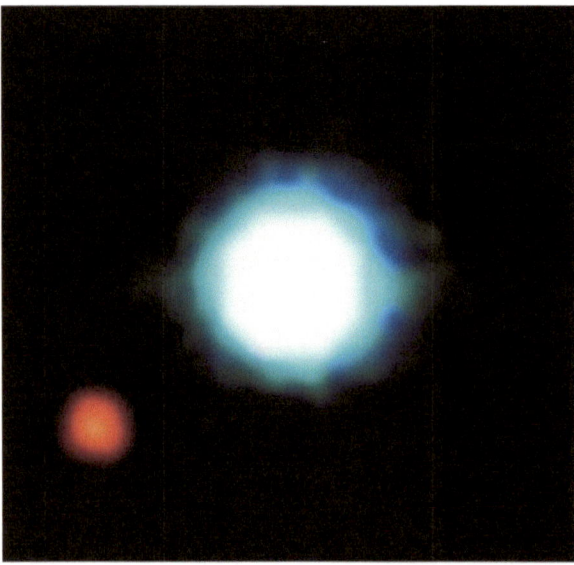

Fig. 6.2 The first direct image of an exoplanet (*red, lower left*), obtained in 2004 at near infrared wavelength with NaCo at the Yepun telescope. 2M1207 b orbits around a brown dwarf star, twice as far away from its star than Neptune from the Sun. Credit: ESO

and that this first image, even though it was indeed the first, was not really what everyone had been hoping for.

The obvious thing to do was to pursue investigations around the hot, young star Beta Pictoris, since even 10 years earlier, the warping of its disk, combined with certain other clues, had suggested that an exoplanet might have been the cause. In 2008, the close collaboration between the groups in Grenoble, Meudon, and ONERA, whose various actors we have encountered so far in this story, finally came to fruition. An extremely hot exoplanet, with a surface temperature of 1500 °C, about 8 times the mass of Jupiter, came out rather clearly on the images, some 8 astronomical units, or one light-hour, from its host star Beta Pic (see Fig. 6.3). Once again, the result was contested and the bets were down. The orbit of Beta Pic b, as it was named according to the prescribed terminology, was in a plane that was seen edge on when viewing from the Earth, and the planet would soon disappear behind the bright light of the star. However, it was due to reappear on the other side of the star in the autumn of 2009, at a date that had already been calculated. And it did indeed

Fig. 6.3 The close environment of the star Beta Pictoris, itself hidden by the corona-graph disc, in near infrared light in a composite image. The dust disk is seen edge-on, as imaged in 1996 with adaptive optics (ADONIS) at La Silla, and shows its warp. In 2008, the exoplanet Beta Pic b shows up as a bright dot in the inner part, imaged by NaCo on the VLT. Its distance from the star is similar to the distance from Saturn to the Sun. Credit: ESO/A.-M. Lagrange et al.

reappear, moving away from the star at first, then moving back toward it. Since that time, its motion has effectively been filmed year by year (Lagrange et al. 2010).

There was intense competition across the Atlantic, sometimes involving close collaborations, and the giant Keck (10 m) and Gemini (8.3 m) telescopes with their adaptive optical systems would soon bring a flood of results. At a distance of 130 light-years from Earth, in the same constellation of Pegasus but otherwise completely unrelated to the star 51 Peg mentioned above, another star known as HR8799 was being observed with these two telescopes by Christian Marois, a Canadian only just in his thirties. He was a student of René Racine, the very same astronomer who had in 1984 already recognised the bright future of adaptive optics, carrying out his first tentative studies on the Canada–France–Hawaii telescope. With a profound sense for what might be achieved, Christian had turned his whole attention to the search for exoplanets and had been accumulating images of this hot and massive star since 2004. From measurements made by the IRAS infrared satellite, it was known to possess a disk containing an abundance of dust and gas. In November 2008, Christian Marois and his group published a superb image showing three massive planets in orbit around the star, while a fourth that would be confirmed later on (see Fig. 6.4) (Marois et al. 2010). Their orbital plane was almost perpendicular to the line of sight from the Earth, as confirmed in 2009 by an image of the disk of debris produced in the far infrared by NASA's

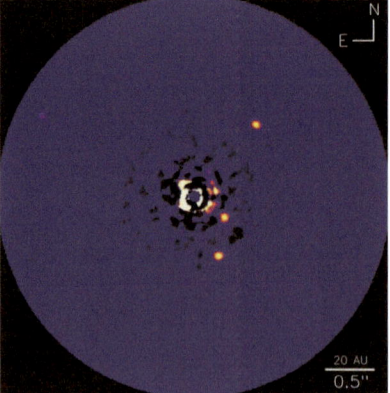

Fig. 6.4 Image of the neighbourhood of the star HR8799 a, obtained with the Keck Telescope in 2010 and confirming the previous discovery. Planets b, c, d and e (e at the field edge on the left) are identified (Marois et al. 2010). The star image is partially masked, leaving some speckles. Scale is given at bottom right. Credit: NRC Canada, Keck Observatory and C. Marois

Spitzer Space Telescope. It was a kind of mini Solar System, viewed from above, containing planets a little more massive than Jupiter, Saturn, and Neptune.

We may leave it to the purists to work out whether Anne or Christian was the one who brought humankind the first image of an exoplanet. What is important is that their work would finally put an end to any reasonable skepticism about the possibility of imaging such objects directly.

For on 13 November 2008 an image was published, taken by the Hubble Space Telescope in visible light, of an exoplanet orbiting the star Fomalhaut A. This star in the southern hemisphere, in the constellation of the Southern Fish (Piscis Austrinus), is actually very close to us, being only 25 light-years away, and is among the brightest stars in the sky. Infrared observations by the IRAS satellite had identified a disk of debris around it, as for Vega and Beta Pictoris. The coronagraph on Hubble had been essential to block out the bright light of the star, thereby revealing the planet Fomalhaut b, about three times the mass of Jupiter. Its motion was subsequently monitored, showing that it orbits a long way from the star (at a separation of about fifteen seconds of arc as we observe it here on Earth) and within a disk of debris.[11]

The giant telescopes set up at the end of the twentieth century, operating at their diffraction limit, quickly got things moving in this new field. The list of such telescopes is long. We have already mentioned the Keck and Gemini North in Hawaii, the Large Binocular Telescope in Arizona, and Gemini South in Chile. Then there was the 8.2 m Japanese telescope Subaru in Hawaii, the 10.4 m Gran Telescopio Canarias by Spain,[12] and the Southern African Large Telescope[13] completing an impressive panoply alongside the VLT. Soon after, a new generation of telescopes, those of the twenty-first century, would go still further, to be seen in Chap. 8. I hardly need to say that all the direct images produced by these telescopes, both the current ones and those to come, would still only reveal these exoplanets as unresolved points. When Galileo observed Jupiter, he saw the planet as a disk, while its four Medicean satellites, known today as the Galilean moons, appeared as points whose orbital motions he could observe hour by hour, as anyone can do today with a good pair of binoculars. We had to wait for Michelson in 1891 to measure the satellite Io as a disk with a resolved diameter of one arc second. How long would we have to wait, and what new level of acuity would be needed, before this tiny point

---

[11] https://en.wikipedia.org/wiki/Fomalhaut and https://en.wikipedia.org/wiki/Fomalhaut_b.

[12] http://www.gtc.iac.es/ and https://en.wikipedia.org/wiki/Gran_Telescopio_Canarias.

[13] Inaugurated in 2005, the SALT has a mirror of diameter roughly 10 m, but it is not circular: see https://www.salt.ac.za and https://en.wikipedia.org/wiki/Southern_African_Large_Telescope.

in orbit around Beta Pic or the four points of light moving around HR8799 would finally reveal some details on their surface?

In 2019, many other images of exoplanets have been obtained, but none have yet been detected as close to their star as Beta Pic b. Moreover, observations provide a great deal of information about the interaction between this planet and the debris disk around the star (Lagrange and Chauvin 2012). In 2013, Anne's group checked out the archived images obtained between 2008 and 2010 by other astronomers using adaptive optics on the Gemini South 8.2 m telescope in Chile. To everyone's surprise, the planet Beta Pic b was indeed there, not completely hidden! In 2010, it had indeed been recorded at a separation of 0.4 seconds of arc from its star, thereby providing independent confirmation of the up to then unique observations made at Paranal (Boccaletti et al. 2013). Christian Marois carried out the same survey of the archives produced by the Hubble Space Telescope and found three of its four planets on an image taken …in 1998! Perhaps we never look closely enough at the observations we have made.

Transits: A Powerful Method

Earlier, I mentioned the European space mission Darwin and its unfortunate destiny. In 1994, another space mission was under evaluation, this time French and going by the name of Corot, spearheaded by Annie Baglin at the Paris Observatory. The idea was to study oscillations in stars, or starquakes, a subject known as asteroseismology, by analogy with the study of earthquakes (seismology) or oscillations of the Sun (helioseismology), which were well established areas of investigation. These oscillations cause tiny periodic variations in the brightness of a star that could be detected by Corot's instruments. In Meudon, back in 1966, Jean Schneider had suggested observing the transit of an exoplanet passing in front of its star, something which should also cause a tiny variation in the apparent brightness of the star, and he had made detailed calculations of what such an event should look like. After the discovery of the exoplanet 51 Pegasi b in 1995, at the *Observatoire de Haute-Provence*, it looked like a good idea to add the study of exoplanets to Corot's main objectives. The decision was taken in 1996 and the Corot mission suitably adjusted to cater for this new task. At our research center in Meudon, Daniel Rouan was given the responsibility for this project.

The idea was simple. If an exoplanet happened to have an orbital plane viewed edgewise from the Earth, each time it passed in front of its star, the light received from that star at the telescope would very slightly decrease during the transit. Since this phenomenon should repeat in a regular way upon each revolution, there could be no risk of confusing it with any other, random variability in the star's brightness. Such a drop in brightness would no doubt be very slight, perhaps a few percent for a large planet like Jupiter, down to a ten thousandth of a percent if the planet were like Earth. However, if all the necessary precautions were taken with regard to stability, such a difference should be detectable. And in a sample of a few thousand stars that might be observed, a few exoplanetary orbits would just by chance turn out to be viewed edge on from our vantage point on Earth. A transit would then occur for the Earth-based observer and the planet's orbital period could thus be determined. Better still, the diameter of the planet could be calculated immediately from the measured drop in luminosity and the diameter of the star (Rouan and Baglin 1998). The latter was generally approximately known by deduction from the star's observed properties, even if it had not been measured.

While the Corot mission was in preparation, David Charbonneau, another young Canadian astronomer, of the same age as Christian Marois, was working at the Harvard College Observatory, the long-established and prestigious research institute in Cambridge, USA, where Wes Traub, already presented above, had designed the IOTA interferometer, then operating in Arizona. So why not try to observe a transit from the Earth's surface? Indeed, if the planet were big enough and if the site were chosen carefully, with a sufficiently uniform atmosphere to avoid scintillation, the periodic drop in luminosity might just be measurable. Now, using the indirect method of measuring periodic variations in velocity (velocimetry) and looking at the results from eleven candidates, the star HD209458 had been identified as possessing a planet that was both massive and close to its host star (less than 0.1 astronomical units), rather like 51 Peg. As the star was very bright, a small telescope would be sufficient. The one used by David Charbonneau had a mirror that measured just 10 cm. In August and September 1999, a transit was regularly observed with a drop in the star's brightness reaching 1.5% each time the planet came in front of the star. The observation of this first transit was a historic moment. The planet was then fully identified as having a mass equal to 0.6 times the mass of Jupiter and an orbital period of 3.5 days (Charbonneau et al. 2000). Being so close to its star, at a distance of just 0.05 astronomical units, meant that HD209458 b fell into the 'hot Jupiter' category, whose surface temperatures would exceed 700 °C. Its radius was determined to be 1.5 times the radius of Jupiter. Indeed, the beauty of the transit method, which soon became an exceptional tool for these studies, was that it could provide the size of the planet directly from the size of the star, which was generally known. In addition, since the mass could be deduced from the orbit, it would also provide the density of the planet, a very important quantity to discriminate between gaseous and rocky planets. Note, however, that it was an indirect method, since the exoplanet was only detected by its capacity to block the light emitted by the star.

Since transits could be observed from the Earth's surface despite the effects of the atmosphere, it was easy to imagine how much better it would be to observe them from space. The Corot observatory gave a perfect demonstration.[14] Launched in December 2006, it operated with great success for almost 8 years, discovering the transits of almost 625 exoplanets, among which at least one resembled the Earth, the exoplanet Corot 7b, announced in 2009 (Léger et al. 2009). NASA launched a similar mission in 2009, called Kepler, which was even more sensitive. We shall have more to say about Kepler's extraordinary crop of exoplanets detected by observation of transits, right up to the end of the mission in 2018.[15] And so the two approaches joined forces: those of Earth-based observation without the blur and the search for transiting exoplanets by spaceborne observatories.

The SPHERE Instrument

Coming back to Earth, let us return to Paranal and the adaptive instrument NaCo. As soon as it went into operation in 2002, Anne was already dreaming of its successor. Indeed, it can easily take a decade to develop and build a new instrument, and during this time ideas will blossom and new techniques will spring up. This successor would be called SPHERE, short for Spectro-Polarimetric High-Contrast Exoplanet Research. Decided in 2004 and implemented in 2014 at one of the focal points of the giant Melipal, SPHERE came about by merging two proposals that were originally in competition, one put forward by Anne's group, the other by Thomas Henning's at the Max Planck Institute for Astronomy in Heidelberg. David Mouillet, the 'discoverer' of the exoplanet Beta Pic 51 b, thanks to the warping of the host star's dust ring, took over from Jean-Luc Beuzit, the mastermind behind NaCo. David would help Anne to develop the new instrument.

Dedicated to the study of exoplanets, each of SPHERE's functions was optimised with this in mind. The instrument had its own very high performance adaptive optics but over a very small field, since the star and planets in the image are generally separated by less than a second of arc. The light that gets

---

[14]https://en.wikipedia.org/wiki/CoRoT.
[15]https://en.wikipedia.org/wiki/Kepler_space_telescope.

deblurred in this way can then be carried to a coronagraph, a spectrograph, or a polarimeter, as required by the observer. The first uses a mask to eliminate most of the light from the star itself, while the second analyses the light from the planet and can for instance be used to investigate the gaseous composition of its atmosphere, if it has one, and the third is sensitive to the polarisation of scattered light, and in particular light scattered by dust grains in the circumstellar disk. Two hundred and sixty nights of viewing with the telescope were allocated by the ESO over several years to the teams that had put together the SPHERE system, in return for their efforts to build an instrument that could be used by the whole astronomical community.

Three years after it went into service,[16] SPHERE had observed some six hundred stars in nearby young star clusters in the southern constellations of Horologium, Pictor, Scorpius, and Centaurus, systematically seeking out the presence of planets. With its state-of-the-art adaptive optics—known as extreme adaptive optics (ExAO)—SPHERE can spot a giant gaseous planet with a mass similar to Jupiter's at a distance of a few astronomical units from a star and up to a few hundred light-years from Earth. This is about a hundred times better than the original ComeOn instrument. One particular statistic clearly emerged from these observations of hundreds of stars: hot Jupiters are relatively rare. Theoretical models would have to explain this conclusion.

Apart from this indispensable systematic survey, SPHERE also helped to establish the relationship between disks and exoplanets, the first being a kind of nursery where the second are born by a formation process that is still far from being established. I shall limit myself to a single beautiful example. At the beginning of this story, in Paranal, I questioned Anne while she was preparing her night of observation with SPHERE. When I asked her what was the most beautiful result to date with this instrument, which she had designed and built with Jean-Luc Beuzit and the eleven co-authors of the paper cited above, she replied: PDS 70 b. But what was that?

Let's go back to the first models of the formation of exoplanets, those that were made when we were still designing the VLT. In the interstellar medium, a cloud of hydrogen gas sprinkled with a little dust collapses under its own gravity. Around the emerging star, a thin flat disk begins to form, with a diameter of a few hundred astronomical units. Within this disk, gravity begins to form clusters of matter called planetesimals, first very small, reaching only a few millimeters, then gradually increasing in size to a meter, a kilometer,

---

[16]SPHERE is described in full detail in Beuzit et al. (2019). Many discoveries and major contributions are mentioned.

and so on, bringing together gas and dust covered with ice and continuing to rotate about the star with the rest of the disk. These lumps gradually fatten themselves up, consuming the matter that remains in the disk. At least, this was one of the scenarios, among many still hypothetical variants, which could lead to the formation of a giant gaseous planet, or indeed a less massive telluric planet, each having fed upon the matter in the disk by accretion. Such a disk is known as a transition disk because in the long run it will have practically disappeared once the planets are fully formed. This is the case in our own Solar System, where the large rocks wandering through space between Mars and Jupiter are doubtless the residue.

So how could one find out whether that corresponds to reality? In 2014, the ALMA interferometer, operating in the submillimeter wavelength range, produced a magnificent image of the disk around the young star HL Tau (see Fig. 6.5). This was the disk that François Roddier had investigated using his instrument Pue'o. With the title *Revolutionary ALMA image reveals planetary*

Fig. 6.5 The disc surrounding the star HL Tauri, as observed in 2014 near 1 mm wavelength by the ALMA millimetric interferometer in Chile. The image size is 1.8 × 1.8 arcsec and the angular resolution is 20 milliarcsec. The gaps visible in the dust and gas disk are due to planetary formation, the material being used up to form the planets. Credit: ESO

genesis, the observation used the maximal interferometer baseline of 15 km at an altitude of 5100 m in the Chilean Andes to reach an angular resolution of 20 milliseconds of arc.[17] The thermal emission from the gas and dust showed not only the disk, viewed from an angle of about 45° to the plane of the disk, but clear gaps that would seem to correspond rather well to the scenario just described. The genesis of the planets, which remain invisible to the radiotelescope, can be considered responsible for clearing out these gaps, where they have been gathering up the gas and dust.

Four years later, in 2018, SPHERE took another great step forward when viewing the disk of the star V1032 Centauri, also called PDS 70. In 2004, the disk was identified on a near-infrared image made from observations at the VLT using NaCo. Other high resolution images would follow, in particular those produced by ALMA. They showed a three-quarter view of a disk measuring only 100 milliseconds of arc across, with an inner region clear of emission and an outer boundary. The characteristics of the emitted light showed that the smallest dust grains had all disappeared at a certain distance from the star. And soon the search was on to find a planet that might have done that. The SPHERE image, taken in the near infrared (see Fig. 1.5), showed a gap about 54 AU across, and a planet was indeed detected there, some 200 milliseconds of arc from the star (Keppler et al. 2018). In a second paper published at the same time, the authors gave the characteristics of this young planet, which had a similar mass to Jupiter and was estimated to be about 5 million years old. Its orbital motion was also established (Müller et al. 2018).

Another object was attracting a great deal of attention, in particular from Anne and her colleagues using SPHERE. This was Proxima Centauri b, discovered in 2016 by the indirect velocimetric method, using the 3.6 m telescope at La Silla (see Fig. 6.6).[18] This is the closest exoplanet to Earth, at just 4.2 light-years. Some even dream of actually sending a camera there to take close-ups! There are no limits to one's ambitions in the quest for detail! Proxima Centauri b is very close to its star, which is cool enough for this planet to be located in the so-called habitable zone. Would it be possible for SPHERE to obtain a direct image, and perhaps even a spectrum showing the composition of the planet's atmosphere? Now that the question had been raised, we should soon get an answer (Lovis et al. 2017).

---

[17] https://www.almaobservatory.org/en/press-release/revolutionary-alma-image-reveals-planetary-genesis/.
[18] https://en.wikipedia.org/wiki/Proxima_Centauri_b.

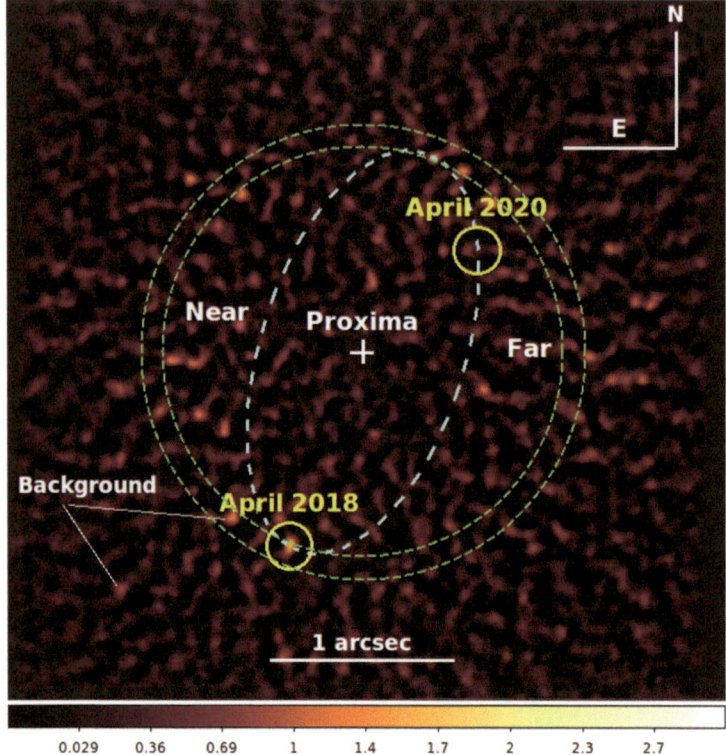

Fig. 6.6 Proxima Centauri, the closest star to the Sun (4.5 light-years), hosts several planets. Planet b was discovered in 2016, and planet c was hinted at, then confirmed in 2020 by several methods: radial velocity, analysis of 4 years of images given by the VLT's SPHERE instrument and analysis of Hubble Space Telescope images. The image here is from SPHERE data (2016 and 2018). Two backround stars are visible, just above the speckle residual noise (scale at bottom), as is the planet. The estimated orbit of Proxima c is in dotted blue, with near and far sides from Earth. Dotted green lines delimitate the search domain. Source: Gratton et al. (2020)

A Stroll Among the Exoplanets

Only 20 years after the discovery of 1995, preceded by the rather atypical discovery of a planet going round a pulsar in 1992, more than a thousand scientists are involved in research on this subject, which barely existed only half a century ago, when attempts to observe the zigzags of Barnard's star had come to nothing. Today, we know that exoplanets are common in our own Galaxy, which makes it highly probable that they will be equally common in the billions of other galaxies that resemble it. This is a fascinating prospect for research and dizzying for humankind. It would seem that there will be one or more low mass planets—the so-called telluric planets, those that have similar masses to the Earth, Mars, Venus, or Mercury—orbiting around the vast majority of the stars we see in the sky. A lesser fraction of stars would also be host to gaseous, hence more massive stars.

Just as this is going to press, in the summer of 2020, a grand total of 4271 exoplanets have been identified with certainty in our galaxy.[19] Most of the stars hosting such planets are less than a thousand light-years from Earth, since these planets become more and more difficult to detect as we search further away. Almost half of these exoplanets were discovered or confirmed by velocimetry, the original, indirect method which led to the discovery of 51 Peg b, while transits, another indirect method used by the spaceborne observatories Corot and Kepler, account for more than a quarter.[20] Moreover, more than a hundred such planets are now directly accessible by imaging, the blur having been vanquished by the methods developed by Anne or by Christian Marois. In addition to the identifications of exoplanets, there has also been detailed study, revealing a wealth of detail, of the disks that gave rise to them or within which they are still in the process of forming, like those of HL Tau, Beta Pic, PDS 70, and AU Microscopii, to name but four encountered in this story.

Another important task is to classify these objects as their numbers increase, taking into account their rather surprising diversity, barely illustrated by the few exoplanets we have described so far in the above. At best, we can explain some of the criteria that have been devised to do this.[21]

The first criterion is the distance from the planet to the star, which is so decisive in determining what details of the exoplanet can be made out. We have already mentioned exoplanets like 51 Peg b which are twenty times closer to their host star (0.05 AU) than the Earth is to the Sun. Fomalhaut b is one of the giant planets situated hundreds of astronomical units away from its star—compare with Neptune, only 30 AU from the Sun.

A second defining characteristic is the planet's mass. Observations are clearly much easier for giant gaseous planets and also for those that are at a certain distance from their star, since the observation is not then swamped by light

---

[19]The number changes almost every week. The number 4271 was reached on 18 June 2020. These statistics are kept up to date at the website of the Paris Observatory already mentioned above, along with graphs and analysis: https://exoplanet.eu/diagrams/.

[20]An informative, elementary visual simulation of the discoveries made by NASA's Kepler mission is available under the name Kepler Orrery V. An orrery is a clockwork model revealing the dynamics of the Solar System, named after the fourth Earl of Orrery, for whom one was made: https://www.youtube.com/watch?v=Td_YeAdygJE.

[21]A recent summary, among many others, is Lagrange and Léna (2019). I thank my co-author and the publisher for allowing me to use this source here.

from the star. Low mass telluric planets, like Mars in our own Solar System, are difficult to observe. A broad range has nevertheless been covered, from the mass of the Earth—about one three-hundredth the mass of Jupiter—to thirteen times the mass of Jupiter in the case of the most massive giants. Among these are planets with masses intermediate between those of the Earth and Jupiter, which have no equivalent in our own Solar System.

Some categories have emerged that are still rather ill-defined, such as 'Jupiters', 'Saturns', 'Neptunes', and 'Earths', by analogy with the planets in the Solar System. Other categories bring together 'massive Jupiters', 'super-Jupiters', 'mini-Jupiters', 'mini-Neptunes', 'super-Earths', and 'mini-Earths', depending on the mass or the radius of the given planet. The transit observed by David Charbonneau was made by one of the 'hot Jupiters', because it was suitably massive and close to its host star. There are also 'hot Neptunes', depending on the distance from the star, and 'gas giants', 'ice giants', and 'ocean planets',[22] the latter referring to their composition or their surface state. The first rather rudimentary models for the formation of planetary systems, those developed over the 1970s to the 1990s which we mentioned above, when the VLT was being designed, had to be revisited in order to explain all these rather surprising properties.

Several hundred multiple planetary systems, that is, containing at least two planets, have been identified around stars of a similar age to the Sun, or older. A good example is provided by the four planets orbiting the star HR 8799, which is 60 million years old, later to be joined by another, all discovered by Christian Marois, as described above. None of the multiple systems already detected around stars like the Sun actually resemble our own Solar System.

No detail could be made out on the direct images of these exoplanets, which just appeared as bright points, more or less spread out by what's left of the seeing and the spreading due to diffraction. Referring back to Antoine Labeyrie's earlier prediction, we were still a long way from that remote time when we would be able to make out the street lamps in their towns, or even see their forests! However, it will be perfectly possible to learn more about them, and in particular about their atmospheres. Any chemical element present in the atmosphere of the planet will tend to emit or absorb light at certain characteristic wavelengths. So by observing the spectrum of the bright point in the image, provided that it is well separated from the host star, we can deduce the presence and also the relative abundances of a whole range of

---

[22] Alain Léger, who was one of the designers of the Darwin mission with Jean-Marie Mariotti, introduced the expression 'ocean planet'. These are less dense than the telluric planets, with a slightly larger radius, and whose surface is completely covered with liquid water, a few tens of kilometers deep.

atoms and molecules. In this way, it has been possible to identify carbon, oxygen, sodium, iron, magnesium, water (H_2O), carbon monoxide (CO), carbon dioxide (CO_2), and methane (CH_4).

Here is one remarkably subtle application. This idea, as clever as it is simple, was put forward by the young Dutch astronomer Jens Hoeijmakers, then working in Berne and Geneva. Imagine an image of Beta Pictoris produced by the SPHERE instrument on the VLT, which has been deblurred as far as possible and on which one can make out the faint spot corresponding to the planet, while the brightness of the star has been removed by the coronagraph. Completely removed? Well, not quite, because neither adaptive optics nor the coronagraph can do a perfect job. Here and there in the image there will be faint speckles resulting from these imperfections, due to this residual light from the star. These speckles could easily swamp the signal from the planet, especially when it comes closer to the star as we view it. If we can isolate the light that is characteristic of the carbon monoxide molecule (CO), absent in the light from the star but present in the light from the planet, these speckles will disappear, since produced solely by the star's light, and the planet will suddenly appear in all its splendour—perhaps accompanied by other, less bright planets belonging to the same system. On the other hand, doing the same thing with the light characteristic of methane, no such thing occurs because there is no methane in the atmosphere of Beta Pic b.[23]

But here we must end our brief stroll among these new worlds, although it hardly does justice to them. There is already so much to say and yet tomorrow there will be so much more than I could recount today! As I attempt to describe the work of so many scientists, sometimes overwhelmed by the depth of their insights, I stand in admiration of the wealth of nature and the intelligence of those who continue its exploration today.

Interferometry and New Worlds

In parallel with indirect methods for detecting exoplanets like velocimetry and transits, adaptive imaging by very large Earth-based telescopes is thus a remarkable tool for the direct detection of these planets, but also the matter surrounding them, from which they have been born. So what about our hopes and expectations for optical interferometry? Were they vain, despite the much

---

[23] *En chasse de molécules pour trouver de nouvelles planètes*, Communication of the Geneva Observatory, 19 June 2018. https://www.unige.ch/communication/communiques/files/3415/2932/9180/Hunting_molecules_to_find_new_planets.pdf.

greater resolution it could deliver? Now that the VLTI was going full steam ahead, using both the giants and the auxiliary telescopes, it was worth re-examining the question.

I mentioned earlier that, when we finally emerged from the crisis in 1996, the ESO's PRIMA project was on the table. This project, which proposed to use dual field interferometry at the VLTI for astrometric purposes, was potentially interesting for the study of exoplanets. To understand why, let us recall the old research technique devised by the astronomer van de Kamp in the twentieth century. The idea was to detect regular zigzags on the image of a star. These would result from the presence of a massive planet whose gravitational attraction on the star would impose a zigzag motion on the star's trajectory through galactic space. Such a search did not work for the star Proxima Centauri, but it was still a good idea. The project PRIMA took it up again, with an image quality and astrometric accuracy at least a hundred times better than those allowed by the seeing that had thwarted the efforts of van de Kamp. Unfortunately, during the 2000s, when Andreas Glindemann and then Françoise were directing the project, it turned out that the egg-cup auxiliary telescopes would not be accurate enough to be used with PRIMA, owing to the presence of tiny residual vibrations. This programme, which had seemed so promising, had to be abandoned. However, just the fact of investigating the possibility in Paranal had sorted out some of the basic principles of differential interferometric measurement, showing that it could even go beyond the diffraction limit of the interferometer in the observation of exoplanets. So these efforts were not wasted. The idea of dual field interferometry (fringe tracking) already mentioned will be discussed in detail in the next chapter, since it conditions the performance of the GRAVITY instrument used to study black holes.

Sylvestre Lacour, associated with the large group working on the GRAVITY project, was a student of Guy Perrin, for once again, the motivations described in the present tale were passed down without fail from generation to generation. In 2018, he suggested using the interferometric dual field of the four giant telescopes to observe an exoplanet, something that had not yet been attempted. He chose the star HR8799, which was bright enough for the fringes to be easily detectable. Among its following of four planets, those that had been found by Christian Marois, he chose to inject the light of HR8799 e, at a separation of 0.4 seconds of arc from the star, into the second fibre, and to expose for almost 2 min. The fringes produced by the star were used 'blind' to stabilise those of the planet, which could then be measured. Moreover, an astrometric accuracy of about 100 microseconds of arc was achieved, thereby establishing the distance between star and planet. A spectrum from the planet showed dust clouds in its atmosphere. The paper was entitled *First direct detection of an exoplanet by interferometry. K band spectroscopy of HR8799 e* (Lacour et al. 2019). Compared with images produced by a single telescope corrected for the seeing, the gain in resolution was by a factor of 20, and the gain in astrometric accuracy by another factor of 20![24]

The faintness of an exoplanet observed from the Earth makes it difficult to obtain images by interferometry, but not impossible, as Sylvestre was able to show using a long exposure and taking a spectrum. However, the situation is better for the disks associated with the presence or the formation of a planetary system.[25] With SPHERE on VLT and analogous adaptive instruments on Gemini, Subaru, and Keck, this huge field of investigation has already proven itself, the resolution reaching about twenty milliseconds of arc. The vast majority of stars known today to have planets orbiting around them are less than a few hundred light-years from Earth. At these distances, a resolution of 40 milliseconds of arc can isolate details measuring 4 astronomical units in the near infrared and 40 astronomical units in the mid-infrared. Getting a more detailed look depends on using interferometry, which can gain a factor of 20 on these values, going from the diameter of one of the giants (8 m) to the longest baseline that can be reached by the egg-cups, i.e., 200 m. The formation process of the planetesimal progenitors of the planets, the evolution of the gaps they clear out on their orbits, and the fate of the disk of dust and gas are all issues that can only be addressed with resolutions of the order of one astronomical unit. No wonder then that the instrument MATISSE, designed to detect cold dust in the mid-infrared, was awaited with such impatience (Kobus et al. 2019).

The interferometric study of dust disks is of course interesting in itself, but the statistical data from these disks is also of great importance for systematic exoplanetary search programmes, now that we know that there are billions of them in the Galaxy. Indeed, if there is too much dust in the disk, the light it scatters from the star, or indeed, at longer wavelengths, its thermal infrared emission, may well completely swamp the faint light reflected by the planet, especially when one is looking for telluric planets. This is why, in 2014, the PIONIER instrument measured the fringes produced by the four egg-cup auxiliaries of the VLTI. Stéphane Ertel, who was in Anne's team in

---

[24]Note added in proof. The exoplanet Beta Pictoris c has been discovered by the indirect radial velocity method in 2019. In 2020, this planet is directly observed with Gravity (Nowak et al. 2020). This is the first example of the convergence of the two methods, each bringing precious information.

[25]The site http://w.astro.berkeley.edu/~kalas/disksite/pages/links.html is a valuable source of information about transition disks and other disks. In particular, it contains simulators showing how these disks evolve.

Grenoble, and Olivier Absil, a Belgian colleague who had started out with us in Meudon, began a systematic study of images of possible dust disks around a hundred or so stars, including Beta Pic (Ertel et al. 2019). With a resolution of 10 milliseconds of arc, the structure of the disks with a light emission of around a thousandth of the emission from the host star could be perfectly distinguished from the star itself. This was a fine demonstration of the potential of interferometry for studying disks! And it was a kind of revenge for Olivier Absil, who had been involved in the abandoned European space project Darwin. Even before his involvement in the Darwin space mission, Olivier had already been working on this exo-zodiacal infrared light.[26]

Since these were indeed 'other worlds', these two examples show that the VLT interferometer did indeed fulfill its promise as we had been hoping back in the 1980s, at a time when we knew almost nothing about exoplanets. And this is only a beginning.

In addition to interferometry, we may mention another method of exploration. At the beginning of Chap. 6, we discussed the failed search for zigzags in the observed trajectory of Barnard's star, such as would be caused by the presence of a planet orbiting around it. At the time astrometry did not consist in any strict sense in reducing the blur in an image filled with detail, but only in determining with the greatest accuracy possible the position and any change in the position of such a star relative to other stars. Using an Earth-based telescope and making repeated measurements many times over, the maximal gain in accuracy, obtained through improved statistics on successive images, is by a factor of ten to fifty relative to the seeing disk. But a telescope in space can do better. The HIPPARCOS mission launched by the European Space Agency in 1989 had already gained a first factor of 20, determining the positions of thousands of stars with an accuracy of one millisecond of arc. With this impetus, a second European mission called GAIA, launched in 2013, gained another factor of a hundred, at least for the brightest stars, thereby reaching positioning accuracies of the order of 10 microseconds of arc. GAIA observed more than a billion stars whose zigzags, if they have any, can now be carefully monitored.

---

[26]So named by analogy with the phenomenon of zodiacal light in our own Solar System. I myself had carried out some investigations, albeit superficially, of these dusts, source of zodiacal light in our own Solar system, aboard the Concorde-001 aircraft during the African solar eclipse of 1973. See Léna (2015).

7

Our Neighbour, the Black Hole

While I have been telling the tale of the VLT, and since we have witnessed earlier in this story the arrival on the scene of active galactic nuclei (AGN), accompanied discreetly by speculations about black holes, I hope the reader has been awaiting this chapter with bated breath. During the ensuing 20 years, attention has become more and more focused on the likely presence of a massive black hole at the center of the Galaxy. With the efforts of Reinhard Genzel, Frank Eisenhauer, Guy Perrin, and many others, this still mysterious object gradually assumed an important place in the preparation of the interferometer, its construction, and its final implementation. Here then is the grand finale which I was hinting at when we met at the top of Paranal in the first pages of this book. But in order to fully appreciate the achievement and before going into the details, I must begin by describing the historical context of black holes.

At the beginning of my story, I discussed the remarkable idea put forward by the Reverend John Michell in England at the end of the eighteenth century: how would it be if light itself could be imprisoned by the gravitational attraction of a massive body? In those days, that supreme achievement of the human mind, Isaac Newton's theory of motion, known today as classical mechanics, was the absolute reference. Newton had formulated his laws of motion in mathematical form, starting with his law of inertia and his famous relation between force and acceleration. These laws could be used to calculate the trajectory of any moving body with great accuracy as soon as one knew the forces acting on it. The motions of the stars and practical problems of motion on the Earth all came under the rule of these same laws.

© Springer Nature Switzerland AG 2020
P. Léna, *Astronomy's Quest for Sharp Images*, Astronomers' Universe,
https://doi.org/10.1007/978-3-030-55811-6_7

Newton had also understood that an object launched into space with sufficient speed—the so-called escape velocity, equal to 11 km/s at the surface of the Earth—would be able to tear itself away completely from our planet's sphere of gravitational influence and move away indefinitely, eventually reaching the remotest confines of the Solar System and beyond.

In Michell's day, the speed of light was already known to be about 300,000 km/s, and his question was as follows: could there be a body massive enough and dense enough that the escape velocity at its surface would be equal to this value? By a simple calculation, Michell worked out the mass this hypothetical body would have to have, assuming that it had the same density as the Sun, and he found quite correctly that it would have to have a radius some 500 times the Sun's. He concluded that, in that case, no light, and hence also no massive object, would be able to get away from the surface of the body with this mass. For light as for matter, this body would behave as a gravitational well. Soon afterwards, Pierre-Simon de Laplace in France asked a similar question.[1]

Published in 1905, Albert Einstein's special theory of relativity profoundly modified this Newtonian conception of the relationship between light, mass, space, and time (see Fig. 7.1). For one thing, space and time become indissociably interwoven, and the trajectories of light or massive objects now had to be identified within the structure of this spacetime. However, this magnificent construction did not take gravity into account, something that bothered Einstein in his search for a profound unity in the description of nature. But then, in 1911, returning to a question he had already formulated in 1908, Albert Einstein published *On the influence of gravitation on the propagation of light* (Einstein 1911). It was indeed the problem raised by Michell and Laplace. This question would occupy Einstein, together with his friends Marcel Grossman and Michele Besso with whom he maintained a steady correspondence, until his famous publications in November 1915.

In four papers, Einstein thus extended the special theory of 1905 to his general theory of relativity, which included Newton's gravity, in a completely new conception of motion. At the heart of the new theory, which is used to describe the motions of massive particles in space, are Einstein's equations, more complicated than Newton's. These mathematical relations effectively determine the trajectories of massive objects, and also light, in any space

---

[1] "Due to its attraction, a light-emitting body of the same density as the Earth and with a diameter 250 times greater than that of the Sun would not allow any of its light rays to reach us. It is thus possible that the largest light-emitting bodies in the Universe may, for this reason, actually be invisible." Pierre Simon de Laplace in *Exposition du Système du Monde*, 1796.

Fig. 7.1 Portrait of Albert Einstein when he visited Paris in 1921, painted by Max Wulfart. Two lines are hand-written in German by Einstein: Gut gemalt in kurzer Frist. Das Modell zufrieden ist. Credit: Académie des sciences

containing matter. In Princeton University in the United States, John Wheeler was working on the consequences of the general theory of relativity, exploring the possibilities for black holes and multiple universes. This was how he expressed the intimate relationship between matter—understand 'mass'—and spacetime: "Spacetime tells matter how to move; matter tells spacetime how to curve." Of course, Newton's insights were not completely swept away. Although less general, Newton's theory can still be used to describe many situations frequently encountered in the world around us, and with great accuracy: when the speed of a body is much smaller than the speed of light c, or when the energy of a body associated with the gravitational field it is subject to, e.g., the energy acquired by a stone of mass m falling to Earth, is very small in comparison to the mass energy of the body (mc^2).

This is not the place to explain how, between 1915 and the present time, Einstein's general relativity has become one of the pillars of modern physics, along with quantum mechanics.[2] We shall limit ourselves to a brief excursion into the world of black holes, explored with exceptional acuity by the GRAVITY instrument at the top of Paranal. So let's get ready.

---

[2]See the books by P. Binetruy, G. Chardin, and J.-P. Luminet cited in the bibliography.

What Is a Black Hole?

One exceptional figure stands out among all the others: the German astrophysicist Karl Schwarzschild, already mentioned above. While working for his doctorate, which he began in Strasbourg (then German) and continued in Munich, he became fascinated by the interference fringes obtained by Albert Michelson in 1891. He applied Hippolyte Fizeau's method to many double stars. When the general theory of relativity was published in 1915, Schwarzschild was the first to use Einstein's equations to calculate the structure of spacetime around a massive non-rotating body whose light emissions would be unable to escape it (Schwarzschild 1916). He showed that this object, which nobody yet called a 'black hole', corresponded to a singularity in spacetime. At this singular point, the values of physical quantities tend to infinity, something that physics is no longer able to represent. For this reason, Einstein never quite came to terms with the singularity.

Through his calculations, Schwarzschild showed that there is a certain length called the Schwarzschild radius, which depends only on the mass M of the object, Newton's universal gravitational constant G,[3] and the speed of light c. In fact, this radius, which is a consequence of Einstein's equations, is simply proportional to the mass M of the object. It is interesting to note that the calculation using Newton's theory gives exactly the same result. Quantitatively, the differences between the two theories are often very slight. Inserting the mass of the Earth or the Sun into the formula for the Schwarzschild radius, we obtain the values 1 cm and 3 km, respectively, provided we neglect the rotation of these bodies, since this does modify the results somewhat. So what is the significance of this radius? It is in fact the radius of a sphere, centered on the singularity and called the event horizon, with the property that any light that happens to be within that horizon will never be able to get out. The calculation of this length provides the first truly physical characterisation of the black hole. Schwarzschild died in May 1916 from an illness contracted on the Russian front where he was a lieutenant in the artillery. Who knows what he might have achieved if he had survived? And the same can be said for so many who fell during this absurd war.

But the debate over these strange objects was far from over. Here is a little anthology of remarks by the great cosmologists: "The singularity in Schwarzschild's field is thus fictional" (Georges Lemaître, 1933); "I think there

---

[3]The constant G determines the acceleration due to gravity g at the surface of the Earth in terms of the mass of the Earth. The value is $g = 9.81\ \mathrm{m/s^2}$. This acceleration is in turn what determines the weight W of a body of mass m through the relation $W = mg$.

should be a law of Nature to prevent a star from behaving in this absurd way" (Arthur Eddington, 1935); "The 'Schwarzschild singularity' does not appear for the reason that matter cannot be concentrated arbitrarily" (Albert Einstein, 1939).[4] However, during these early years, these same physicists were making tremendous theoretical progress in understanding the nuclear origin of the energy radiated by the Sun and stars, and hence beginning to understand also the structure, diversity, and evolution of stars. These investigations were themselves based upon the developing possibilities for spectroscopic observation of the stars which could be used to determine their temperature, mass, composition, and gravitational field. Extremely dense stars made entirely of neutrons were hypothesised and their existence confirmed with the discovery of so-called pulsars in 1968, some time after the Second World War, which had inevitably slowed down research in the field of astronomy. But picking up on Einstein's question quoted above, could there be a mass concentration that was even denser than the matter in the core of a neutron star?

In 1968, John Wheeler suggested using the name 'black hole' for these objects, which still only enjoyed a theoretical existence. Then, in 1969, using Einstein's equations, the mathematician Roger Penrose showed that any collection of matter collapsing within the Schwarzschild radius would go on collapsing to the singularity. But at the time, all that was just a fascinating theory.

It is no easy matter to imagine what a black hole would be like, and the same is true of other objects postulated by physics today, such as atoms and quarks. Even light, so often discussed in this tale, turns out to be rather like the bat in the fable,[5] being sometimes wave, sometimes particle. Our mental categories, which represent and name 'things' in the world, result from the long evolution that has produced our brains and our senses, and are conditioned by this. But nature, for its part, is in no way restricted by the cognitive limitations of our brains and their limited ability to represent the world. Rather surprisingly, mathematics comes to our aid here. Its 'unreasonable effectiveness'[6] somehow

---

[4]Quoted by Thibaut Damour. In *Les trous noirs: leur nature, et leur rôle en physique et astrophysique*, Académie des sciences, February 2018. See http://www.academie-sciences.fr/fr/Colloquesconferences-et-debats/trous-noirs-nature-et-role.html.

[5]"I am a bird: see my wings …What makes a bird? The plumage, of course. I am a mouse." J. de La Fontaine, *The Bat and the Weasels*.

[6]It was the theoretical physicist Eugene Wigner (1902–1995) who spoke of the "unreasonable effectiveness of mathematics". He was of Hungarian origin but naturalised American, winning the Nobel Prize for Physics in 1963.

manages to inform us of the laws of the Universe.[7] But how could one talk about black holes using everyday words and language?

Following the contemporary physicist Thibault Damour, we might try this: "a tube in spacetime bounded by a horizon (its surface), localised in space and persisting in time, at the surface of which a bubble of light is running on the spot and within which spacetime is torn and crushed". Or perhaps, returning to more familiar physical characteristics, a black hole is entirely characterised by its mass (in kg), any rotation it may have, measured by its angular momentum, if it rotates like a spinning top or the Earth, for example, and finally, any net electric charge it may carry. Each of these assertions can of course be questioned and should as far as possible be tested by observation. For example, Stephen Hawking raised the question of the supposedly eternal stability of a black hole, i.e., the question of whether it must indeed 'persist in time' forever. Indeed, he showed that a black hole of stellar mass or greater must eventually disappear, gradually evaporating by what was then called Hawking radiation over an extraordinarily long period of time, longer than the fifteen billion or so years presently considered to be the age of the Universe. For low mass black holes, such as may have existed in the distant past of the Universe, the time they would require to evaporate by this process would be considerably shorter.

Are There Black Holes in the Universe?

After this extraordinary theoretical work, pushing human understanding so far beyond our familiar representations of the world, it remained only to actually discover these objects and see whether such pure products of physicists' deductive logic and intrepid imagination really could exist in the world. But given that neither matter nor light can escape from such an entity, would astronomical observation be able to reveal their existence through their effects on their surroundings? Whatever else, it was a fine research programme, beginning in 1965 and still far from completion in 2019.

Fig. 7.2 Artist's vision of the black hole Cygnus-X1, discovered in 1964. The binary system includes a supergiant star (*left*) and a black hole, orbiting each other in 5 days. The gravitational field of the black hole, about 10 times the mass of the Sun, sucks the gas from the star and creates an accretion disk at very high temperature, source of the X-ray radiation. Credit: ESA

Stellar Black Holes

In Chap. 1, I mentioned the first reasonably certain evidence for the existence of a black hole in our own galaxy. This was in 1965, when X-ray astronomy came into being, pioneered by Riccardo Giacconi, future Director General of the ESO, among others. Using an X-ray telescope carried aboard a rocket, he discovered a bright X-ray source in the constellation of Cygnus, which he called Cygnus X-1 (see Fig. 7.2). The discovery was confirmed between 1965 and 1973. Situated in our galaxy and with such strong X-ray emission, Cygnus X-1 cried out for an explanation, and one was duly found. It went as follows. A system comprising two stars held together by gravity, in other words a binary star system, had evolved over time. One of the stars was more massive and collapsed into an extremely dense, invisible object which sucked hydrogen from the other star, as could be observed directly with a telescope. When matter falls in this way it is called gravitational accretion. A considerable amount of

---

[7]See in the bibliography R. Omnès, *La révélation des lois de la physique*, O. Jacob, 2008.

gravitational energy, of the same kind as revealed by the accelerating fall of a stone, is thereby transferred to the hydrogen atoms. The hydrogen gas heats up due to increasingly energetic collisions between the atoms and when it reaches a high enough temperature, it radiates X-ray light.

Since this first discovery, about twenty other such systems have been identified in the Galaxy and in neighbouring galaxies through their X-ray emissions. These are good black hole candidates. They result from the collapse of a star with a mass only slightly greater than the mass of the Sun. As this star belongs to a binary system, matter will always be accreted from one component onto the one that has now collapsed, and this explains the intense X-ray emissions we observe.

Supermassive Black Holes

There is another category of black holes. These are referred to as supermassive black holes because they can have masses up to several billion times the mass of the Sun. The existence of such monsters, which seems highly improbable at first glance, has actually been suspected since the 1960s. I have already introduced them above. Two very intense radio sources were known at the time, 3C273 and 3C48, both associated with very low emissions of visible light. When observed from the Earth, these powerful radio sources looked like geometric points and could not be resolved by telescopes. Their spectra were highly redshifted. When interpreted in terms of the expanding Universe, such redshifts showed that they were extremely remote in time, and hence also in space. Such radio sources were called quasi-stellar sources or quasars, because they did in fact look almost like stars.

It was soon discovered that quasars were always associated with a galaxy. Gradually, during the 1970s and 1980s, observation revealed the existence of a large class of galaxies called active galaxies. The intense radiation from their central regions indicated an internal source of energy, called the active galactic nucleus (AGN). These sources were so powerful that they could only be attributed to gravitational accretion onto an extraordinarily massive object. It was thus suggested in 1964 that a black hole would provide the best possible explanation for the intense radiation, despite the widespread skepticism about these objects, which remained highly speculative.

The small region from which the radiation is emitted—not just X-ray, but also visible, infrared, and radio—is only about the size of the Solar System, while the energy radiated from it is equal to some thousand billion times the radiation from the Sun. This energy results directly from the expression for the equivalence of mass and energy, so well known in the form $E = mc^2$. If a mass m is totally annihilated, an energy E becomes available, here in the form of radiated light. Hence, the maximal power that can be radiated by an active galactic nucleus comes directly from the amount of infalling mass, eventually totally annihilated there, per unit time. This maximal power is called the Eddington limit. The existence of such a limit is due to the pressure exerted by the emitted light on the infalling matter. When the luminosity is intense enough, it can actually stop the matter from falling. Active galaxies emit a prodigious level of power, referred to by astrophysicists as the 'luminosity', equal to a significant fraction of the Eddington limit. In fact, AGNs are extraordinarily efficient at converting gravitational energy into light. Depending on the galaxy, the attractive mass of the nucleus can be anywhere between a few million and several billion times the mass of the Sun.

Today, more than 200,000 quasars have been identified. The general consensus is that they signal the presence of supermassive black holes accreting matter supplied by the host galaxy. The distances of these host galaxies makes it well nigh impossible to see any detail in the process. However, it is possible to make out a disk of gas rotating at a considerable distance from the black hole, and also a jet of matter expelled to huge distances along the axis of rotation of the disk. It is generally accepted that most galaxies have a massive black hole at their center which would have formed by accretion throughout the long history of the galaxy. It is not always possible to observe the environments of these black holes at galactic centers because the accretion does not always occur in a uniform manner over billions of years. For an Earth-based observer today, there are therefore many 'silent galactic nuclei', while others may be the source of intense light emission. The center of our own galaxy is one of the rather silent ones at the present time.

Regarding the more active ones, we have mentioned them here because of the considerable interest they raised at the time when the VLT was being designed. Indeed, it should come as no surprise that astronomers were quick to react to the idea of obtaining sharper images of AGNs, whether in radiofrequencies or the infrared. We have already discussed the adaptive instrument PUEO (the Hawaiian owl), built by François Roddier for the Canada–France–Hawaii telescope on Mauna Kea. Using this instrument, Daniel Rouan and Olivier Lai produced the first image of the active galaxy NGC1068 (also known as Messier 87), and a superb one it was, some ten times sharper than the seeing limit (see Fig. 7.3).

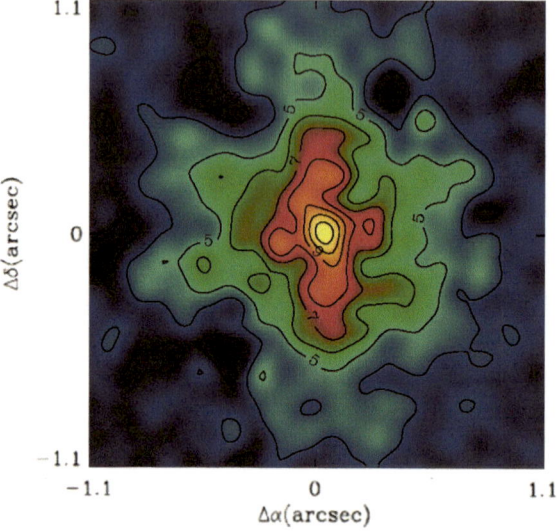

Fig. 7.3   The active nucleus of the galaxy NGC1068 is imaged in the near infrared using PUEO adaptive optics on the 3.6-m Canada–France–Hawaii telescope in Hawaii. The seeing blur is reduced and the diffraction limit of the telescope is reached (0.12 arcsec). The bright unresolved nucleus shows up, as well as an extended structure around it. Image size is 2.2 x 2.2 arc sec. An uncorrected seeing disc would occupy about a quarter of the field. Source: Rouan et al. (1998)

Gravitational Waves and Black Holes

Karl Schwarzschild's black holes were not the only thing made possible by Albert Einstein's general theory of relativity. It also predicted the existence of a new kind of waves, quite different from light but propagating in empty space with the same speed c. These were gravitational waves, which are extraordinarily difficult to detect. Unless perhaps they are emitted by black holes! Needless to say, from the 1950s, physicists began to look for ways they might detect these waves and thereby throw some light on the black holes that might be emitting them.

On 3 October 2017, the Nobel Prize in Physics was awarded to three physicists in the United States, Rayner Weiss, Barry C. Barish, and Kip S. Thorne, for "decisive contributions to the LIGO detector and the observation of gravitational waves", an observation published in February 2016, just 18 months earlier. A week before, on 27 September, two French physicists, Alain Brillet and Thibault Damour, were award the gold medal of the CNRS, Brillet for his experimental work on Europe's VIRGO interferometer and Damour for his theoretical work. Their contributions played a key role in the detection of gravitational waves. The prestige of these awards, the fact that they occurred at the same time, and the fact that the Nobel prize was attributed so quickly when it is generally only decided after a lengthy confirmation process, shows just how important this discovery was in the history of physics and astrophysics, coming a century after the prediction had been made from the general theory of relativity.[8]

This incredible achievement had been made possible by using Michelson interferometers! However, the idea was not to equip a telescope in Fizeau's manner, but to measure a length with an unbelievably high level of relative accuracy, better than one part in a thousand billion! To do this, the LIGO and VIRGO interferometers apply exactly the same principle as the instrument conceived by Michelson during his stay in Potsdam. I mentioned this instrument in Chap. 4 when discussing the key experiment used to lay the ghost of the aether, which had up to then been considered necessary for the propagation of light in a vacuum. Albert Michelson himself had been awarded the 1907 Nobel Prize in Physics for this beautiful experiment and the new understanding of light it necessitated. I won't go into the description of these modern gravitational wave detectors LIGO and VIRGO, as this would take us too far from our main theme. Let us just note what was concluded from these first observations. In 2020, there are already a number of additional detections, and there will be more in the future, thanks to the detectors now operating in Europe, the United States, India, and Japan.

It is very common to find pairs of stars bound together by their mutual gravitational force, forming a binary system in which each revolves around their common center of mass. It may then happen that one of these stars has evolved to become a black hole. Accretion in such a system then turns it into an X-ray source of the kind described above. It may also happen that both stars have collapsed to become black holes. Like their progenitors, these two black holes will revolve one about the other. Rather analogously to the energy lost by the motion of electrical charges in an aerial, which is carried away by a radio wave, this system of two black holes will radiate gravitational waves into the surrounding space in a way described by Einstein's equations. The system thus loses the energy carried away by these waves, while the distance between the two black holes must decrease to compensate for it. In a fraction of a second, this process will end by the merging of the two objects to form a single black hole. The energy radiated in the form of gravitational waves in this final phase

---

[8]The event was celebrated on 5 April 2016 at the *Académie des sciences* in Paris by a series of talks. These can be accessed on http://www.academie-sciences.fr/fr/Colloques-conferences-et-debats/ondesgravitationnelles-et-coalescence-de-trous-noirs.html/.

is tremendous. When it reaches Earth, it will cause the distance between the mirrors of the interferometer to oscillate ever so slightly, but at a level that is just detectable today. The form of this oscillation accords with what would be expected from the coalescence of two black holes, as described by general relativity.

The first historic detection of such an event was thus made on 14 September 2015, and subsequently named GW150914. The detected wave was soon attributed to the merger of two black holes with masses 30 and 36 times the mass of the Sun, slightly more massive than the conventional stellar black holes. The huge energy E carried by the wave was quite simply equal to the energy obtained by annihilating three times the mass M_S of the Sun, i.e., $E = 3M_Sc^2$. The wave extended into space like the circular ripple on the surface of water. After travelling 1.3 billion light-years, it reached the Earth and shook spacetime locally, as revealed by measuring the relative lengths of the interferometer arms (Blanchet 2017). As explained by Kip Thorne, one of the Nobel prizewinners mentioned above, the power released in an instant of time in the form of gravitational waves is greater than the total power radiated by all the stars in the visible Universe!

This phenomenon, which has since been observed on several occasions, is without doubt the best direct proof we have to date of the reality of black holes, in this case, low mass black holes. However, a still more reliable demonstration must await a very precise measurement of the detected signal to check that it corresponds exactly to the calculations based on Einstein's equations.

The Center of Our Galaxy: The Great Enigma

On a beautiful summer's night at our northern latitudes, the luminous glow of the Milky Way rises up from the southern horizon. Myriad stars in the constellation of Sagittarius shine brightly between vast dark regions where not a single star is visible. Our Galaxy is like a huge flattened disk, almost circular in shape, whose structure is maintained by its rotation and the force of gravity, the latter binding together the hundred or so billion stars of the Milky Way. Our Sun is placed somewhere toward the outside, but in the plane of the disk, and all the stars, including the Sun, are rotating slowly around the center, the galactic center. Seen from the Earth, the center lies in the direction of Sagittarius. It is hidden by a vast accumulation of interstellar clouds made up of dust and gas, which mark out the dark regions so clearly visible to the naked eye. Who would not be intrigued by this? What might be hidden behind those clouds?

Fortunately, at radio and infrared wavelengths, such dusts become transparent to light. Telescopes detecting these wavelengths can see through the dark clouds and form images of the objects concealed behind them. At a distance of twenty-five thousand light-years, the center of our galaxy is a hundred times closer to the Earth than the center of the Andromeda galaxy (Messier 31), and much, much closer than many other galactic nuclei, hundreds or thousands of times further again. Observing the center of the Galaxy, even a telescope limited by the seeing can view very small regions. It would never resolve details of the same size if it were observing Messier 31. So one of the main programmes of the past half century has been to identify the physical phenomena hidden in the center of our Galaxy by seeking the maximal acuity in images produced at these wavelengths. At least a hundred astrophysicists on both sides of the Atlantic, maybe more, were thus involved in lifting this veil of mystery. And needless to say, this great scientific adventure depended heavily on winning the war on blur.

We should say what exactly we mean when we talk about the center of the Galaxy. Since the hundred billion stars of the galactic disk are engaged in circular motion, the center is precisely the point of space about which this rotation takes place. Naturally, there is no material axis playing the role of a wheel hub. This kinematic point is located in a small region measuring a few tens of light-years across, which would subtend a few minutes of arc at the Earth-based viewer. This is something we know because we have measured the motions of thousands of stars. So what is there in this tiny region? It has been carefully studied and turns out to be full of clouds of hot ionised hydrogen, but also cold molecular hydrogen (H_2) and cold dust clouds. Hot young stars are evolving there, with complex structures and motions, immersed in magnetic fields. These first details were obtained by observations made at radiofrequencies since 1954, then in the far infrared from a stratospheric balloon in 1969. When it was still rather poorly defined, this region was given the name SgrA, where Sgr is the official abbreviation for the constellation of Sagittarius. Twenty years later, in 1974, radio waves indicated the presence of an extremely high energy source of radiation. Its intensity could not be explained by simple heating up of gases and its true nature remained a mystery for 10 years.

First Radiofrequency Images

In 1932, the physicist Karl Jansky, a radio wave specialist involved in building large radio dishes, discovered an extraterrestrial source of these waves and determined its position in the sky to be in the middle of the constellation of Sagittarius. This historic event marked the birth of radioastronomy and the beginning of interest in this region of the sky. Much later, in 1985, five radioastronomers joined forces and linked up six radiotelescopes, spread right across the United States, to obtain a more detailed image of the region at centimeter wavelengths. They thus effectively put together an interferometer in the same way as Ryle had done in Cambridge, but with baselines in the thousands of kilometers.[9] Their observations allowed them to confirm the existence of a unique source in the Galaxy, but resembling most closely the compact radio sources (quasars) at the center of other galaxies. They determined that this source was contained in a tiny region, measuring at most 20 astronomical units, or just three light-hours. As the authors put it themselves, their observations were "best explained by a single massive collapsed object at the galactic centre." (Lo et al. 1985).

The maximal size estimated at the time was just twice the distance between Earth and the planet Saturn. In this same issue of the British journal Nature, the editor John Maddox stated with great confidence that the latest measurements left no reasonable option but to conclude that there was a black hole at the center of the Galaxy. So it looked as though there would soon be confirmation of the hypothesis put forward in 1971 by the British astronomer D. Lynden-Bell and the young Martin Rees, whom we met in Chap. 4 when they encouraged us in our plan to declare war on blur. Even an inactive galaxy like our own can accommodate a black hole at its center, although a much quieter one than the active galactic nuclei of the quasars. From this point, this remarkable region became known as SgrA*. The reader may consult the table at the end of the book for certain technical details regarding the perception of distances and angles on the sky, and also the anti-blur capabilities of the telescopes that have been used in successive investigations of the source SgrA* and the black hole that is probably concealed there.

All attention now turned to this tiny region SgrA*, which could be observed in the infrared and radio, the only wavelengths able to reach us through the clouds of dust and gas. Very long baseline interferometry at centimeter wavelengths had reached its limiting resolution. Infrared could pick up from there, just as the first infrared-sensitive cameras came on the scene in the United States, able to form images at these wavelengths. One particularly bright infrared source called IRS16 looked as though it was a star. Could this

---

[9]This mode of operation of radiofrequency interferometry is known as very long baseline interferometry (VLBI).

be the source SgrA*? And did the center of rotation of the Galaxy coincide exactly with SgrA*, or perhaps with IRS16, or was it somewhere else entirely? Reinhard Genzel had not completely left Charlie Townes and Berkeley, where he remained a professor. However, he was by then spending a considerable amount of his time and the resources of the Max Planck Institute in Garching, right next to the ESO headquarters, on the investigation of the galactic center and the possibility of tracking down the black hole whose presence seemed ever more likely.

The Advent of Infrared Imaging

In Europe in the 1980s, we were busy designing the VLT, and we were well aware of its potential for making discoveries in the infrared. I have already described how, with François Lacombe in Meudon, we had been implementing the very first infrared imagers, which the French Ministry of Defence had made available to us. The mystery of the galactic center was of great interest to me, even though my own understanding of the physics of this turbulent region was nothing compared to the young Reinhard's. However, we had at least to try, and to do that, we had to develop our infrared imaging capabilities.

Our modest camera Rodrigue[10] had 64 pixels. We set it up on a telescope of diameter 2.2 m at La Silla, in Chile, as it was the only ESO observatory at the time. François, who was a first rate observer and a talented experimentalist, would defend his doctoral thesis on infrared observations of the galactic center. Although he did not yet have adaptive optics, which would come 3 years later with the ComeOn system, his ability to select good images among the short speckle exposures, with their varying degrees of speckle, allowed him to reach a resolution of 0.3 seconds of arc, which was exceptional at the time. The image he obtained of the central region was particularly beautiful and I showed it to Lo Woltjer. The latter was so impressed that he decided to put it on the cover of ESO's 1986 annual report. We were very proud of ourselves.

We thus prepared a publication in which our images very carefully localised the stars making up the source IRS16, now resolved into six components, and also the radio source SgrA*. We concluded that the latter was very close but did not appear to coincide with any of the six components, then submitted a

---

[10]The name is perhaps rather surprising. This infrared camera worked in a slightly different way to the CCD cameras already mentioned. Indeed, it was a charge injection device or CID. But from there to imagining its desire for Chimène's eyes, like Rodrigue, the famous hero of Pierre Corneille's play *Le Cid* …!

paper to the main European journal, Astronomy & Astrophysics. The referees made several quite justified criticisms which were easy enough to deal with. But alas! Foolishly, we never actually published our paper. I suppose that, in the autumn of 1987, the political battle I was engaged in with the French authorities to obtain financing for the VLT kept me so busy that I didn't have the peace of mind to work on the publication.

On the other side of the Rhine, in Garching, Reinhard had immediately understood the role that detailed images could play in resolving the mystery. The fall of the Berlin Wall had lessened the tensions of the Cold War and restrictions on supposedly 'sensitive' equipment (from a military standpoint), and he lost no time in importing an infrared camera with 256×256 pixels from the United States. That was already a thousand times more pixels than we could obtain with our rudimentary camera Rodrigue! He called it SHARP and set it up on the 3.58 m New Technology Telescope (NTT) in La Silla. The NTT was the superb telescope designed by the optical engineer Ray Wilson at the ESO in the 1980s to demonstrate a new design of thin, active primary mirror.

The idea of using SHARP was to gain at least to some extent on the blur by selecting the best images among a very large number of short exposures. This technique was inspired by the now standard speckle method devised by Antoine in 1970 and which our team among others had developed in the infrared. Indeed, since the seeing is caused by atmospheric turbulence and is therefore a random phenomenon, certain instantaneous images are better than average. In a sense, they 'fix' those trembling points of the seeing effect that Newton had noticed. By selecting and then superposing them, these can be used to produce a sharper image than would normally be allowed by the seeing effect.

The group in Garching, with Andreas Eckart, thus managed to improve the resolution by a factor of four. It was not yet as good as adaptive optics, but it was better than nothing. In 1993, this group published a paper, illustrated with magnificent images taken in the near infrared and with a resolution of 0.15 seconds of arc (Eckart and Genzel 1996). In the central region around SgrA*, more than 340 sources were identified as stars and their distribution analysed. However, it was still impossible to decide once and for all whether the radio source SgrA* coincided with one of these sources. We were not cited and for a very simple reason: our paper presented 6 years earlier had never been published, and by our own negligence.

A Star Moving Like the Clappers: S2

The images obtained with the camera SHARP, gradually refined over the years, were used by Andreas and Reinhard, with exceptional tenacity, to explore the positions and motions of the stars around SgrA* in ever greater detail. A series of papers were published, each more precise than the last. Meanwhile, in Hawaii, the 10 m Keck telescope had just begun operating. It gathered more light and had better seeing than La Silla. There was stiff competition because a group in California was applying the same image selection method. In 1998, Andrea Ghez, a brilliant young astronomer only just in her thirties who ran the group at the University of California in Los Angeles,[11] published the positions and spectra of 90 stars clustered in the neighbourhood of SgrA*. In this tiny region, they had images obtained with the same selection technique with a resolution of 0.05 seconds of arc, which was a record.

The speeds of the stars were determined by measuring the Doppler effect on each individual spectrum and found to exceed 1400 km/s. The mass of the attracting object, which had to be highly concentrated and able to hold stars in orbit with such high velocities, could be deduced immediately. It was estimated at 2 million times the mass of the Sun. The authors concluded (Ghez et al. 1998):

> Although uncertainties in the measurements mathematically allow for the matter to be distributed over this volume as a cluster, no realistic cluster is physically tenable. Thus, independent of the presence of Sgr A*, the large inferred central density [...] leads us to the conclusion that our Galaxy harbors a massive central black hole.

So in fact, the motions of the stars around SgrA*, already identified by Reinhard in 1996, would perhaps provide the key to the enigma. Each was therefore given a name. Andrea's paper contains a long list in which they are identified by the letter S followed by a number, then the position and velocity at the time of the measurement. And before long, one of these S stars would become famous.

As I have already said, from 1990, adaptive optics was gradually accepted by astronomers and began to produce images on the 3.6 m La Silla telescope. I also described how, in 1996, following the crisis at the ESO, it once again became their priority. The French proposal to build the instrument NAOS, prepared

---

[11]https://en.wikipedia.org/wiki/Andrea_M._Ghez.

by Anne in Grenoble, was finally accepted. Then, in 2000, the instrument NaCo installed on the telescope Yepun, combining the French adaptive optics (Na) and the camera built in Heidelberg (Co), received its first light, at exactly the same time as the adaptive optics on the Keck did the same in Hawaii. And so it was that the rest of the story of the black hole would be written thanks to the discoveries of the deblurred giant telescopes which became available at the beginning of the twenty-first century.

It was not long before things got moving. In the autumn of 2002 an article with the title *A star in a 15.2-year orbit around the supermassive black hole at the center of the Milky Way* appeared in the journal Nature (Schödel et al. 2002). This was a fine convergence of effort after 20 years' work. The authors included a good few among those already presented in the above: on the German side of the Rhine, Reinhard and Andreas Eckart, now working in Cologne, and on the French side, François, Daniel, Anne, and Gérard Rousset, the astronomers who designed and built NAOS,[12] accompanied by Eric Gendron, who built the infrared analyser in 1996. And then of course, at the ESO, the two who orchestrated the whole project, namely, Norbert Hubin on the adaptive optics and Alan Moorwood on the infrared. A magnificent image of the region around SgrA* was obtained in the near infrared using NAOS on Yepun. The resolution was 56 milliseconds of arc, the ultimate diffraction limit of the telescope Yepun and ten to twenty times better than the seeing.

Among the stars in the neighbourhood of SgrA*, S2 is the star (in the other sense!). Its elliptical trajectory had been followed by the camera SHARP since 1992, then by the Keck telescope. It looked as though SgrA* sat almost precisely at one of the foci of this ellipse (see Fig. 7.4). NAOS first imaged S2 in the spring of 2002 and tracked it month by month when it moved closer to this focus, in fact, to within one light-day. This point of the orbit, commonly called the perihelion for the Earth in its orbit around the Sun, and the periastron for a binary star system, is now known as a peribothron in the case of a black hole, from the Greek word *bothros* meaning 'well'–here, in the sense of a gravitational potential well. The orbital period of S2 was determined as 15.2 years. To within a few months due to the remaining uncertainty in the measurements, it should return to its original position around 2018. The mass of the central object has been confirmed as close to four million times

---

[12]Note that the engineers, such as Pascal Puget and others in this case, never appear among the authors for these papers, and yet without them, NAOS would never have existed. This may seem surprising, but there is a kind of silent agreement in research centers that papers of an astrophysical nature will be signed by astronomers, while those of a technological nature, equally innovative and just as essential, are signed by the engineers. However, their role is crucial at every stage and I am not sure that this rather traditional way of sharing the 'glory' can really be justified.

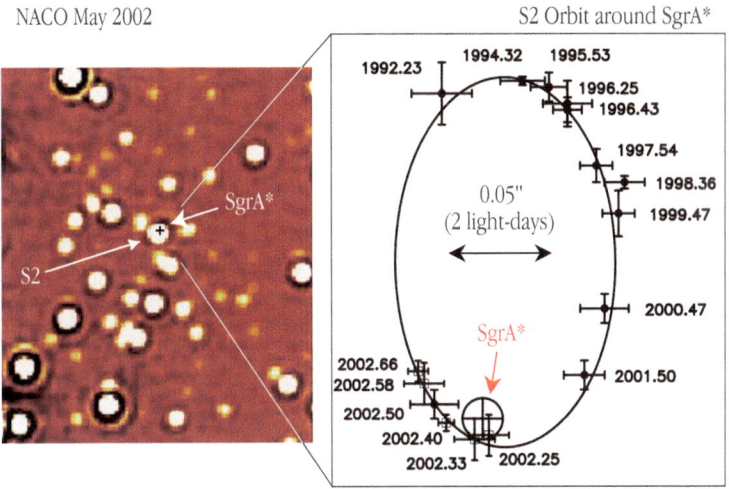

NACO May 2002

S2 Orbit around SgrA*

Fig. 7.4 The beautiful image of the galactic center area (*left*), obtained with the NaCo adaptive optics and camera on the Yepun VLT telescope. Each diffraction-limited star image shows its first Airy ring. At *right*, the reconstructed motion of the S2 star over 10 years. The *circle* shows the position of SgrA*, at one focus of the elliptic trajectory. Credit: ESO

the mass of the Sun. The trajectory followed by S2 deviates in no way from an ellipse. This shows that this star is moving through empty space, without encountering any matter, even when it approaches to within one light-day of SgrA*. The matter attracting it must therefore be highly concentrated in the vicinity of the latter. The editor of Nature made the following comment: "The Milky Way, like other galaxies, is thought to harbour a black hole at its centre. The remarkable observation of a star in close orbit around the Galactic Centre is the first firm evidence that this is so." He then added: "Their success shows the true power of a relatively new observing tool, adaptive optics imaging."[13] And like many good Britons, looking as always toward the United States, he praised this achievement without saying a word about Europe and our NAOS system, mentioning only the Keck telescope in Hawaii!

In 2003, the team in Garching, under Reinhard's lead, brought together all the acquired data in a paper whose title alone said everything about our triumph over blur: *Stellar dynamics in the central arcsec of our Galaxy* (Schödel et al. 2003). Within the blurred seeing disk obtained with Yepun, which spread over one second of arc, adaptive optics made it possible to plot a detailed map

---

[13]Comment in the same issue of Nature, p. 614.

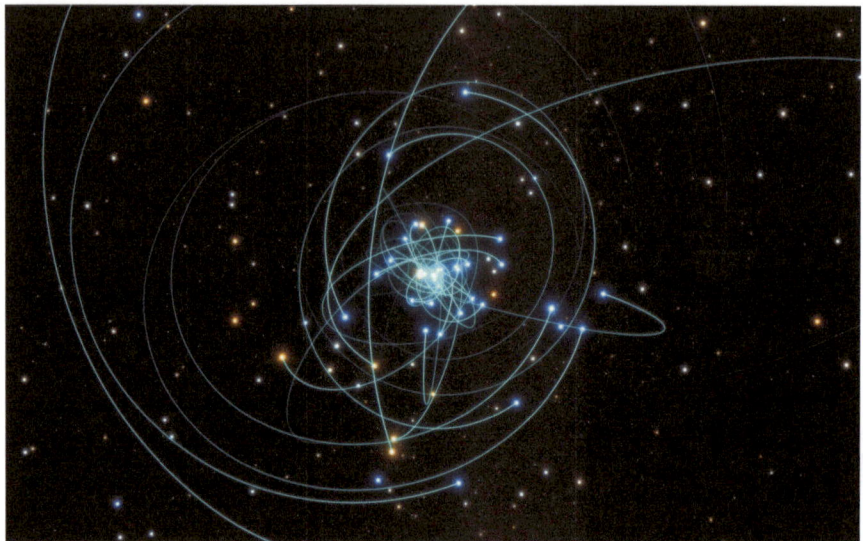

Fig. 7.5 A beautiful three-dimensional ballet of stars, in elliptical orbits around the galactic center black hole. This computer-made image (2018) is based on the measurements of star positions over more than 20 years. The star S2 is one of the innermost. See Gravity Collaboration (2018). Credit: ESO/L. Calçada/spaceengine.org

showing the positions and motions of more than 40 stars, including S2 (see Fig. 7.5). Alternatives to SgrA* being a true black hole were getting scarcer by the day.

Surprising Flares of Light

In the spring of the same year 2003, very close to SgrA*, Reinhard and Daniel observed bright flares of infrared light emission, each lasting a few tens of minutes, using the NaCo camera in Paranal (Genzel et al. 2003). Almost at the same time, a similar observation was made by Andrea Ghez using the Keck telescope. Yet another element in the black hole story!

A few months later, it was not without emotion that I went on my last observing mission to Chile before going into retirement at the university. Alone in the Paranal night, at the controls of NaCo and having asked for Yepun to be pointed toward the zenith where a splendid Milky Way was passing slowly by, a bright point suddenly lit up on the control screen in the image taken around SgrA*. It gradually brightened to a maximum then slowly disappeared over a period of about 10 min. This rare and beautiful gift from the skies marked the end of an era for me. A page was turning, one which 40 years earlier had been almost blank. At this precise moment, the blur had at last disappeared from the screen.

The flares were not easy to interpret, although it seemed likely that these short, bright flashes of light were produced by electrons moving through the strong magnetic fields of the disk surrounding the hole (Narayan 2003). When this kind of phenomenon is produced in such a strong gravitational field, astronomers find themselves with a perfect laboratory to test the predictions made by the general theory of relativity. And this laboratory happens to be rather close to us, relatively speaking, because it's only about twenty thousand light-years away. One thing would be to find out why this particular black hole is so quiet compared with those at the center of active galaxies, since it is only emitting something like a billionth of the Eddington luminosity. At this point in the story, let us jump forward 15 years to find S2 on its elliptical orbit, during its next passage at the peribothron, its closest approach to SgrA*, in May 2018. By then, the new instrument GRAVITY had been set up at Paranal. This would be the first time that adaptive optics and interferometry would work together. At last, these two methods, both present when the VLT first came under the spotlight, would be able to join forces.

The GRAVITY Instrument

In the first chapter of the book, which took place at the top of Paranal, Guy Perrin and Frank Eisenhauer were observing the neighbourhood of SgrA* with the four giants. These sent the light they had gathered into the GRAVITY instrument, which had been masterminded by Frank and Guy with the teams in their respective research centers. This enormous instrument was absolutely unique, contained in a cylindrical steel tank more than 2 m long, in which the temperature was lowered to almost $-200\,°C$ to improve stability and cool the infrared light detectors (Eisenhauer et al. 2011). Installed in the underground laboratory of the VLTI, it received the light simultaneously from the four giants, or from the four egg-cup auxiliaries, depending which were being operated in interferometry.

GRAVITY was designed and built between 2006 and 2015, partly in Germany and partly in France. After receiving its first light in November 2015 and undergoing a little fine-tuning, it began its systematic observations of the galactic center the following year, employing all the sensitivity that can be achieved with four 8.2 m mirrors, and its interferometric resolution far exceeded that of NaCo. In May, it began to map the elliptical orbit of the star

S2. Moving at more than 1000 km/s, S2 was coming close to the peribothron, the point of the ellipse where its distance from SgrA* would be minimal. This point was reached in 2018.

The genesis of the GRAVITY instrument is a great story in itself and certainly deserves to be told before we begin to outline its latest achievements. By 2004, the black hole had been completely cornered. The adaptive optics in Paranal, as in Hawaii, guaranteed the sharpest possible images allowed by the size of the Yepun and Keck mirrors. The VLTI had become operational, as two of the giants had been coupled since 2003 and two of the auxiliaries had also just joined forces. Naturally, we were all dreaming of observing the region around SgrA* and the motion of the stars orbiting it using all four giants together to form an interferometer. The detail in the images would be improved to previously unknown levels, reaching the millisecond of arc or perhaps even better. We would at last satisfy our wildest aspirations of the 1980s.

In order to check the quality of their work without complacency, to plan their future, and to situate themselves in the best possible way in the international context, major research centers like the Paris Observatory or the Max Planck Institute for Extraterrestrial Physics (MPE) co-directed by Reinhard (see Fig. 7.6) in Garching always appoint a visiting committee, usually foreigners in the given country. I myself was asked to join such a committee, which met every 2 years in Garching and which submitted its conclusions to the President of the Max Planck Society, the largest and most

Fig. 7.6 Thirty-five years investigating the depths of SgrA*. Reinhard Genzel, after an observing run on Gravity instrument at Paranal. Reinhard Genzel is co-laureate of the 2020 Nobel prize in physics, for his tenacious work on the SgrA* black hole. Credit : Max Planck Institute für extraterrestrische Physik

prestigious public research institution in Germany, on which the MPE fully depends. After a whole day of hearings, the conclusions of the committee were presented rather ceremoniously and subsequently used to guide the allocation of resources to the institute. Even the directors, who were scientists of international standing, awaited with some trepidation the verdict that would be announced in a snowbound Bavaria at the end of November 2004. Both their funding and their authority depended on it. And the verdict was clear enough: the results obtained concerning the galactic center were considered to "place the institute at the forefront of one of the most remarkable of current developments in astrophysics". And yet, during the presentations, no mention was made of the future interferometric instrument, even though it would be designed over the next few months!

At the time, Thibaut Paumard was on a post-doctoral scholarship at this institute. During his doctorate in Paris, this young Frenchman had grown interested in the stars populating a broad region around the center of the Galaxy, observing them with a spectrograph installed on the Canada–France–Hawaii telescope on Mauna Kea.[14] Rigorous, determined, and an excellent physicist, he was ready now to join Reinhard's group for a few years and complete his training as a research scientist. Frank, Guy, Thibaut, Andreas Eckart, and myself (the young pensioner), came together more and more often at the beginning of 2005. The big question was whether we would we be able to design an interferometric instrument that could observe SgrA* more closely and better than anything that had been used so far, at a reasonable cost and on an acceptable schedule. Although the challenge was great, we decided to give it a go. The result would be well worth it.

In Chap. 5, I explained how, in 2004, following the successful interferometric coupling of the egg-cups, then the giants, we drew up an ambitious plan with the ESO for developing the VLTI over the coming decade. The committee was chaired by Guy and included Reinhard, Frank, Thibaut, but also Andreas Glindemann and Françoise for the ESO. The conclusion of the committee's report stressed the potential of the VLTI for studying the immediate environment of SgrA*. Guy, who was at the time particularly interested in active galactic nuclei, raised a critical point: how could one improve the capabilities of the interferometer to observe much fainter sources? The idea seemed to me so relevant for the study of SgrA* that I immediately

---

[14] Simons and Maillard (1996). Thibaut used a rather special kind of device called a Fourier transform spectrometer, named after the same Joseph Fourier already mentioned. This particular instrument, called BEAR, was designed by the French astronomer Jean-Pierre Maillard, who spent his whole professional life perfecting this kind of system, ideal now for observing the galactic center and a good many other objects.

phoned Reinhard. An appointment was made in Garching and, as a kind of New Year's gift, we all met up at the Max Planck Institute. A few hours later, the main lines of the future instrument had already been laid down!

A few months later, during the 2005 ESO workshop, where the future of the VLTI was discussed on behalf of our three institutes, in Garching, Meudon, and Cologne, Frank presented the idea behind the system which had been called GRAVITY.[15] The proposal, which I shall outline briefly, shows that our dream was realisable. Here it is! When the light reaches us from the sky, its path emerges from the primary mirror of each telescope and is sent underground, where it is reflected by an adaptive mirror (MACAO), before passing through the delay lines of the 'metro' and arriving at last in the underground laboratory. However, the field of view allowed by this long path, which we call the interferometric field, is limited to 2 tiny seconds of arc around the target point. Recall that Jacques Beckers had planned with great foresight to have a wider field, of 8 seconds of arc. However, in 1996, the need to cut costs when the interferometer project was reinstated had limited the size and cost of the moveable mirrors carried by the 'metro', where the light would be reflected to make up for the different arrival times of the given wave front at the different telescopes. This explains why the interferometric field of view was so drastically reduced.

Quite by accident it turned out that the region around SgrA* provided two pieces of good luck! The first was the existence of an infrared source called IRS7, bright enough in the infrared to provide an adaptive correction signal to the deformable mirrors of each MACAO, but too far away from SgrA*, at a separation of 5.5 seconds of arc, to be included in the interferometer field of view. At the focal point of each giant, a specially designed device thus picked up the light from this star and sent it to an analyser which could then calculate and transmit the correction signal to each MACAO adaptive mirror. The star IRS7 was close enough to SgrA* to ensure that the correction would be adequate and that the same quality would be obtained in 'straightening out' the wave for each of these two sources.

The second piece of good fortune was this. Another star, slightly less bright and going by the name of IRS16C, was situated very close to SgrA*, at a separation of only 1.23 seconds of arc.[16] Therefore, this star and the target SgrA*, where the black hole is very likely situated, both fall well within the

---

[15] Eisenhauer (2006). The perfectly suited name GRAVITY was suggested by Andreas Glindemann, who ran the VLTI programme at the ESO.

[16] There is actually a second bright star, called IRSNW, which is close enough to be used, and is in fact used. For clarity in our description, we shall only refer to the star IRS16C.

interferometric field of view. Their respective light waves are thus easily picked up by each of the four giants, corrected for the seeing by the corresponding MACAO adaptive mirror, then advanced or delayed as required by one of the four 'metro' lines. After this long journey, they are received in the underground room and ready to go on into the GRAVITY instrument.

It is amusing to note that, 30 years earlier, in 1986, I had already noticed these two pieces of good fortune and suggested to two of my students[17] that they might examine the environment of SgrA* for their Master's thesis, taking into account the constraints imposed by adaptive optics, which was just at its inception. A map of the region, published in their dissertation, clearly showed these two exceptional pieces of good luck, but we could hardly have imagined just how important they would later become!

GRAVITY operates at low temperature and in vacuum (see Fig. 7.7). This protects its mirrors from any form of dust or pollution. At the input and for each beam, two optical fibres are positioned to select IRS16C and SgrA*, for example, these being neighbouring points in the tiny field of 2 arcsec, or the star S2 when it came close enough to SgrA* in 2018.

Then the optical fibres, like tiny pipes guiding the light, divide to create six combinations corresponding to the six different baselines (AB, AC, AD, BC, BD, CD) that can be formed by the four giants A, B, C, and D. The team from Grenoble had acquired much useful experience in the art of using fibres and integrated optics when they developed PIONIER, and this gave invaluable support for Guy, who came with his own experience on the OHANA project in Hawaii. The final result is that the cameras see and measure six systems of infrared fringes for one of the objects in the field, and six systems for the other. They transmit these measurements to the computer, which visualises them for the observer and archives them. SgrA* and S2 are very faint infrared sources. A long exposure is needed, lasting several minutes, before their fringes appear clearly against the background noise. During this exposure, the atmosphere acts randomly on the light as it arrives and scrambles the fringes. Fortunately, the fringes obtained on the star IRS16C are bright enough to be measured very quickly, before being scrambled in this way. Moreover, this star and SgrA* (or S2) are about one second of arc apart, so the light coming from either of them will pass through almost exactly the same agitated atmosphere. This

---

[17]These were two students from the *École polytechnique*. The first was Denis Mourard, who subsequently worked with Antoine Labeyrie on the GI2T on the Calern Plateau, then made a major contribution to equipping the Californian interferometer CHARA with a spectrograph in the visible. The second student was Nicolas Mercouroff, who chose another direction after his Master's degree.

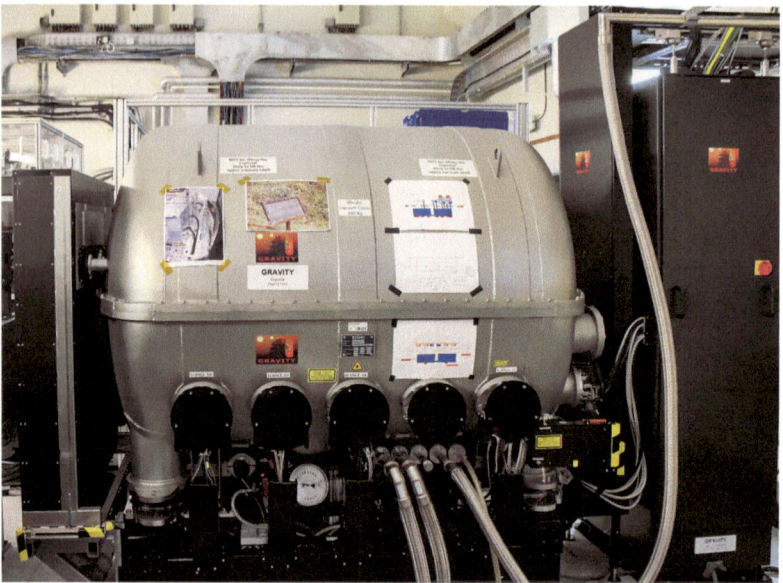

Fig. 7.7 The GRAVITY instrument, installed in the underground interferometric room at Paranal. The huge vacuum tank contains the low-temperature optics and infrared detectors. Four light beams, coming from the four 8.2-m telescopes, enter the instrument through the windows (here with black covers) and recombine inside to form the six sets of fringes. Credit: ESO/G. Rojas

is therefore an ideal situation for applying the dual field or fringe tracking method which gave rise to the PRIMA project, as discussed in Chap. 5.

By measuring the clearly visible set of fringes taken on IRS16C, a position control system can reposition the invisible set of fringes due to SgrA* (or S2) at each instant and hence stabilise it blind, so to speak. So this is how those two successive pieces of good fortune, namely, the two suitable stars just next to the black hole, allowed us to apply the dual field method!

In addition, one must measure the distance from the detector to each of the four primary mirrors with nanometric accuracy, and this throughout the observation. A metrological laser is used. This is much more precise than those now used by architects and surveyors, but operates by the same principle. When he presented this completely innovative project, Frank went into many critical details. He stressed the fact that there was no particular reason why it wouldn't succeed. With great conviction, he described how the previous years of work in the war on blur had prepared our teams to take up the present challenge. He cited the many skills that had been developed, not just at the ESO, but also in European industry, working with optical fibres, integrated optics, infrared analysers, hypersensitive sensors, and laser metrology, which together should allow us to meet the specifications of the present instrument.

At the end of the day, in order to study SgrA*, we had to gain a factor of a thousand on the sensitivity of 2004! Regarding the question of blur in our images, what could be hoped for? Compared with the blur due to diffraction for a diameter of 8.2 m, the maximal baseline of 130 m would already improve the acuity by a factor of about 15, giving a resolution of 3 milliseconds of arc for the infrared wavelength of 2 μm. The possibility of dual observation within a 2 arcsec field of view, never before achieved with such extraordinary accuracy, would allow a relative (hence astrometric) positioning equal to 10 microseconds of arc between the two objects, which is an angle three hundred times smaller than the resolution limit, measured with an exposure of a few minutes. This may indeed seem miraculous, but it really works. Such a favourable situation is well known to physicists: sometimes two quantities cannot be measured with great accuracy, while their difference can, provided that the sources of error in each individual quantity can be treated as similar.

When GRAVITY was finally presented, Thibaut took over from Frank and assigned the following goal: to reach the Schwarzschild radius.[18] He discussed the consequences of Einstein's general theory of relativity that could be put to the test by observations reaching so close to the source of such a strong gravitational field. For recall that this radius, the Schwarzschild radius, is the distance from the singularity within which any matter and any light is irretrievably trapped by the black hole. For a black hole of mass 4 million times the mass of the Sun, it is equal to 0.1 astronomical units, which is one tenth of the distance from the Earth to the Sun, or again, 40 light-seconds. Seen from the Earth at the distance of SgrA*, the corresponding angular size of this radius is just 10 microseconds of arc, a value to be slightly corrected for local spacetime effects. Thibaut's presentation was in perfect agreement with the anti-blur and astrometric ambitions that Frank presented for the GRAVITY instrument!

Of course, things were not going to be quite that simple. We had estimated the cost of this instrument very roughly to be around 4 million euros, which would be shared between the various institutes making the proposal. In the end it would actually cost twice that. Moreover, if it was built, it would use the four

---

[18]T. Paumard, Scientific prospects for the VLTI in the Galactic Center: getting to the Schwarzschild radius. In *The power of optical/IR interferometry*, Springer, 2006. More detail can be found in M. Grould, F. Vincent, T. Paumard, G. Perrin, General relativistic effects on the orbit of the S2 star with GRAVITY, Astr. Ap. **608**, A60 (2017).

giants in Paranal over many nights, thus preventing their use for programmes put forward by other astronomers. Not to mention the fact that the levels of performance required were likely to raise doubts about whether it was realistic. However, within the ESO generally, and in particular at the final review by the ESO Council, the scientific stakes were considered high enough for the project to be accepted without much hesitation. It was as important for fundamental physics as for astrophysics and was based on the expertise of those making the proposal. There are still not that many experimental checks on general relativity and it was certainly worth giving it a go.

GRAVITY was officially approved in 2008 and work got under way. Apart from the environment of SgrA*, the astronomical community was well aware that other targets, from exoplanets to active galactic nuclei, would also benefit from the new instrument, either with the 8.2 m giants or with the moveable 1.2 m egg-cups. To build it, the three bodies originally involved in the proposal were joined by astronomical institutes in Grenoble, Heidelberg, and Portugal. This consortium, bringing together more than a hundred researchers, was called the GRAVITY Collaboration. All papers announcing its future discoveries would then be signed with this name. It took 10 years of intense effort to build GRAVITY, and there was no time to be lost because a competing project had already begun. This was the Event Horizon Telescope, to be discussed below.

In September 2015, the coupling between GRAVITY and the four egg-cup auxiliaries was tested in Paranal. Then, in June 2016, the four giants were connected and the fringes observed from the galactic center.[19] A year later, the interferometric configuration with GRAVITY was judged ready for service (Mérand et al. 2017; Gravity Collaboration 2017).

The venture was a total success (see Fig. 7.8). The position of the star S2 was determined with previously unthinkable accuracy as it approached SgrA*, at an angular separation of 0.05 seconds of arc, which is just 50 milliseconds. Better still, during one of the viewing nights, there was an infrared flare and its position could be determined by its fringes. It coincided precisely with the position of SgrA*. This was therefore the first interferometric detection of infrared light emission from the immediate vicinity of the black hole, and it was no ordinary result for the 134 members of the GRAVITY Collaboration who signed the paper. Many of them have already been mentioned here for their role during the twenty previous years. GRAVITY was indeed ready for service, and the newspaper *Le Monde* paid tribute to it under the headline *GRAVITY, the black*

---

[19]https://www.hq.eso.org/public/news/eso1622/.

Fig. 7.8 Paranal in 2016, happiness and satisfaction: GRAVITY is operational. Standing between the interferometric egg-cup telescopes are (*left to right*) Frank Eisenhauer, the author, Antoine Mérand, and Pierre Kervella. Credit: Sang Shao-Hua

hole hunter,[20] although the journalist hardly mentioned the role played in this achievement by the close cooperation and complementarity that had made it possible across the Rhine, nor the long and impressive story of the war on blur that had led up to GRAVITY's final victory.

Brush with the Schwarzschild Radius

On 19 May 2018 of the terrestrial observer's calendar, the star S2 passed through its peribothron, or pericenter, the point on its orbit closest to SgrA*. This point is situated at about 14 light-hours, or 100 astronomical units, from SgrA*. We were now looking at things on the scale of our own Solar System, since the dwarf planet Pluto is about 50 AU from the Sun. Viewed from the Earth, the angular distance separating S2 from SgrA* on the sky is 12 thousandths of a second of arc (12 mas) when it passes through its pericenter. On the other hand, the pericenter is still far from SgrA* in the sense that

[20] *GRAVITY: le chasseur de trou noir*, Le Monde, 10 May 2016.

it is still at a distance of 1400 times the Schwarzschild radius of the black hole. Between March and June of that year, over fifteen nights of the Chilean winter, the exchanges were intense between Europe and Chile, as astronomers spared no effort, each keen to contribute in his or her area of expertise. Frank, Reinhard, and Guy were there to coordinate the dialogue. The position of S2 was measured relative to SgrA*, reaching the incredible astrometric accuracy of 30 microseconds of arc. This was thirty thousand times better than the seeing limit, which, at the beginning of my story, seemed insurmountable! In addition, on certain nights, the spectrum of S2 was measured using the instrument SINFONI set up on the telescope Yepun. This last measurement was essential to completely determine the velocity of S2 using the Doppler effect.[21]

Here we recover the benefits, which may seem paradoxical on the face of things, of the dual field obtained using two fibres. The resolution of VLTI, determined by the diffraction limit, was 3 milliseconds of arc, but the differential astrometric accuracy between SgrA* and S2 was a hundred times better thanks to the possibility of making this detailed comparison between the two sets of fringes, one from each source of light.

Recall that GRAVITY was built to check the physical effects predicted by general relativity when the gravitational field is very strong, as in the immediate vicinity of a black hole. So it is no surprise that the title of the first paper, which came out in the European journal Astronomy & Astrophysics in September, was entitled *Detection of the gravitational redshift in the orbit of the star S2 near the Galactic center massive black hole* (Gravity Collaboration 2018). Apart from the classic Doppler effect due to low speed motions, spacetime itself is affected by the very strong gravity here, which produces an additional spectral shift that can be calculated. The measurements made on S2 agree with the predictions of general relativity here. They imply that the massive object causing this strong gravitational field has a mass of 4.1 million times the mass of the Sun, making it ever more likely to be a black hole.

A second paper described a still more extraordinary achievement, one that had been contemplated by the members of the collaboration since 2011 (Vincent et al. 2011). During the viewing nights, three flares were observed. Interferometer observation had lasted long enough for each to be able to determine their position and motion over the few tens of minutes that the flare

---

[21] Seen from the Earth, the velocity of S2 can be decomposed into two components: one in the plane of the sky, which is what is determined by measuring its positions on a nightly basis, and the component perpendicular to that, along the line of sight, derived from the velocity, itself measured by using the Doppler effect.

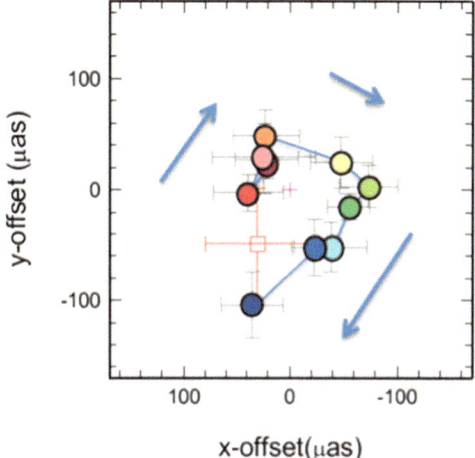

Fig. 7.9   Differential astrometry reaches as close as possible to the supermassive black hole. The GRAVITY instrument detects a bright flare and follows its motion for 30 min (*dots from brown to blue*, to identify time passing). *Arrows* point to the nearly circular motion around a central point (*black cross*). Resolution reaches 50 microseconds of arcs (μarcsec) (Gravity Collaboration 2018).

was visible. And surprise! The positions of the three flares, so-called hotspots, lay rather precisely on a trajectory close to a circle, and each was moving along it in the same direction. This circle was extremely close to the black hole, since it had a radius estimated at 3.5 times the Schwarzschild radius. Such an orbit would be completed in just 50 min, at a speed equal to one third of the speed of light. Naturally, these numbers had to be interpreted with reference to Einstein's equations, given the strength of the gravitational field in a region so close to SgrA* (see Fig. 7.9).

Now it happens that general relativity makes a quite remarkable prediction. The theory describes the last stable circular orbit that a material body can have around a black hole. This orbit is called the innermost stable circular orbit (ISCO). If the body is anywhere within this orbit, it either gets swallowed up forever or ejected. The radius of this orbit depends on any rotation, or spin, the black hole may have about its own axis. So, although the observations of these three moving flares are not sufficient to draw any definite conclusion about this rotation, these measurements and others that would be made over the coming months and years should allow us to determine this second characterising factor of the black hole, namely its spin, since its mass is now established (Gravity Collaboration 2018).

As the authors pointed out with all the necessary caution, this description on the basis of a single lump of hot matter in orbit, the source of the flares, although justified mathematically, may be an oversimplification of the complex phenomena occurring in the immediate neighbourhood of SgrA*. So we must await further news. But note that no image ever before acquired could show us the motion of matter so very close to a black hole.

Healthy Competition

In any great scientific exploit, it is rare to find that only a single person or a single collaboration is involved. The investigation of our neighbour the black hole is no exception. I have already described how, in 1985, a radiotelescope array in the United States with a maximum baseline of 3000 km had been observing SgrA* by very long baseline interferometry (VLBI) at centimeter wavelengths. Reaching an angular resolution of 2 milliseconds of arc, these measurements had revealed the exceptionally compact nature of this source. Then, 9 years later, using the same method but with a wavelength ten times shorter, hence an acuity ten times greater, the best estimate of the compactness of the source could be reduced by a factor of ten. The size of the source was thus estimated to be less than 200 microseconds of arc as viewed from the Earth (Rogers et al. 1994), and over the following years this number could be reduced to 130 microseconds of arc. But how could we get closer still to the true value? If we could set up an interferometric array with an intercontinental baseline, of the order of 10,000 km, and better still, at a wavelength close to the millimeter, the blur due to diffraction would be reduced by a factor of thirty relative to the initial measurements at 3 cm (the wavelength effect), then by another factor of three (the baseline effect), making a total reduction by a factor of around a hundred. From images with a diffraction blur of 2 milliseconds of arc, this would take us to images a hundred times better, with a blur of just 20 microseconds of arc.

These values were the target of a project called the Event Horizon Telescope. This 'virtual telescope', which would have a size of almost 10,000 km, close to the diameter of the Earth, is actually made up of many radiotelescopes, one of them being situated at the South Pole. These are connected up to work together in aperture synthesis as an interferometric radio array,[22] in the same way as GRAVITY uses the four Paranal giants in concert. The name of this necessarily international project says everything: the aim was to get as close as possible to the event horizon of the black hole. If the black hole is not spinning, its event horizon is situated at the Schwarzschild radius. For the mass of SgrA*,

---

[22]https://eventhorizontelescope.org/.

known to be 4 million times the mass of the Sun, the Schwarzschild radius would be 0.08 astronomical units, which is about 40 light-seconds. But what should an Earth-based observer see? An object, or rather a shadow devoid of all light, a black disk of diameter 0.16 AU, which from this distance would subtend a tiny angle of just 20 microseconds of arc? Not exactly, because the path of light passing close to the black hole is strongly perturbed by the high curvature of spacetime and by the effects of any spin the black hole may have. According to the various assumptions made, the angle would be more like 55 microseconds of arc (Broderick and Loeb 2013).

The resolution of the Event Horizon Telescope, calculated using the longest possible baselines, hence slightly less than the diameter of the Earth, could reach 26 microseconds of arc, sufficient to resolve the horizon. This shows that the name was consistent with the aim. Moreover, at this resolution, the Event Horizon Telescope could produce a genuine image by aperture synthesis. For its part, GRAVITY, a comparatively modest undertaking and much less costly, can achieve the ultimate astrometric accuracy, improving by a factor of two.

The Event Horizon Telescope began work in 2006, but things have proved extraordinarily difficult. Since 2017, it has been regularly announced that results were on the way. The great success came in 2019, with an image of the shadow of the supermassive black hole, located at the center of the M87 galaxy. A resolution close to 20 microarcsec was achieved (EHT Collaboration 2019). Regarding the goal to image a similar shadow near the SgrA* black hole, it is not yet reached, although images were obtained in 2019, using only a subset of the EHT antennas (Issaoun et al. 2019). In a sense, GRAVITY won the first round with its astrometry measurements (Fig. 7.9), but the Event Horizon Telescope will soon make real images at a similar resolution.

It would not be possible to end this chapter, in which we have finally got the measure of our neighbour the black hole, without coming back to the great victory that these results demonstrate over the problem of blur. The Earth's atmosphere and its seeing are corrected by adaptive optics of 8 m class telescopes, the diffraction limit is lowered by long baseline interferometry, up to 130 m, and the astrometry is refined using the dual field method. Since 2018, each of the four moveable egg-cup auxiliaries has been equipped with its own adaptive optics system with the nice name NAOMI. It was built by the Institute for Planetary Sciences and Astrophysics in Grenoble, where Anne is currently working, and was set up in 2018, 26 years after the first recommendation was explicitly set out as part of the VLT programme, and almost 40 years after we had with François Roddier understood the importance of combining adaptive optics and interferometry. Astronomy requires a great deal of patience!

All this was the result of a huge joint effort by dedicated astronomers, talented engineers, and companies ready to take significant technical risks, working together on both sides of the Rhine, and of course elsewhere, too, all supported by public institutions who still consider that knowledge for knowledge's sake is a worthwhile human undertaking.

8

The Future Lies in the Details

And we have nights more beautiful than your days—Jean Racine, *Letter XIV to Mademoiselle Vitart*

The explorations of exoplanets and the massive black hole by SPHERE and GRAVITY attest to our final triumph over the blurred image. They have succeeded in going beyond the seeing limit and reaching the limits imposed by diffraction when observing at the VLT with infrared light. However, this is not quite the last word on blur. Hoping that my story has not seemed too long, let us take one last stroll around the summit of Paranal, where each night we gather so many new images full of myriad details.

And then, let us ask: what next? After the successes of the half century that has just gone by, which have so inspired the present, what new incentives will motivate the future? Will the next chapter be written by the supergiant telescopes, then by the hyper-supergiant telescopes, built at the tops of mountains on the Earth? Or perhaps on the plains of the Moon? Or indeed, in space, far removed from any form of atmosphere? Will astronomers still be focused on light waves, or will their attention turn to gravitational waves?

The Most Beautiful Observatory in the World

Anne and Guy were at my side in Paranal at the beginning of this story and they have been present throughout. But, fascinating as exoplanets and black holes may be, they are not the only objects whose secrets astronomers have been

© Springer Nature Switzerland AG 2020
P. Léna, *Astronomy's Quest for Sharp Images*, Astronomers' Universe,
https://doi.org/10.1007/978-3-030-55811-6_8

trying to penetrate for over three hundred nights each year, carefully picking amongst the photons they have sent us from the depths of time. Paranal hosts many observers from the world over, as well as engineers improving spectrographs and detectors, and students there to observe for the first time, not without emotion. I can no longer invite you to visit, but I would encourage you to consider the important roles that certain ideas, those of Babcock and Linnik, or again those of Fizeau, Michelson, and Labeyrie, have come to play at Paranal on a daily basis.

Artificial Stars

Almost every night during 2019, absolutely straight, fine, bright rays of reddish-orange light were emitted by powerful lasers from the giant telescope Yepun at the top of Paranal, disappearing into the dizzying heights of the night sky, scattered with a multitude of stars. How strange, these astronomers, lighting up the sky in this way, to get the best out of their omnipresent adaptive optics systems! But under Norbert's enlightened but demanding management, things have come a long way at the ESO since the resurrection of the war on blur in 1996.

The instrument NaCo has been shifted from Yepun and placed at one of the focal points of Antu, while SPHERE has been set up on Melipal. Among the four giants, the best equipped telescope is now without doubt Yepun. It also has an extraordinary deformable mirror and four powerful lasers mounted on its structure, the whole thing forming what is known as the Adaptive Optics Facility (AOF).[1] Three instruments placed at its various focal points exploit the various ways of reducing the blur for the benefit of European astronomers. One day, perhaps, if finances allow, Antu, Kueyen, and Melipal will also benefit from a similar panoply. All this has been achieved by the hard work of some of the best engineers, scientists, and technicians in the world.

The little deformable mirror at the heart of our instrument ComeOn in 1989 measured about 10 cm and comprised 19 elements for correcting the light wave. At that time, we couldn't have done any better. Today, Yepun has a secondary mirror of diameter 1.12 m, built in France by a company called SAGEM. The idea of such an ambitious deformable mirror was put forward in 1992 by Piero Salinari, an Italian astronomer who seemed to have stepped straight out of a Quattrocento painting, and who worked at the Florence

---

[1] The reader can download *The Adaptive Optics Facility Booklet* from the ESO website at https://eso.org/sci/facilities/develop/ao/images/AOF_Booklet.pdf.

Observatory on the Arcetri Hill. My friend Piero, originally skeptical of the war on blur, finally took up arms in the 1990s, at which point his research center became key collaborator for the ESO. One could expect no less from Italy, in this stunningly beautiful city where modern astronomy was born when Galileo pointed the first refracting telescope towards Jupiter! Yepun's secondary mirror is a technological masterpiece. It is a thin, deformable sheet of glass, in fact, 2 mm thick, positioned right at the top of the telescope structure, where it reflects the light coming from the primary mirror. Its flexible surface is rapidly activated by 1170 electromagnets. This allows it to 'straighten out' the light it receives from the primary mirror and send a beautiful wave front downstream to the instruments. There are three of these, but before describing them, let us just throw a little light on those mysterious laser beams that spring forth from Yepun each night.

Was this strange and brilliant idea actually born in the dark quarters of a United States army that dreamt of star wars under the presidency of Ronald Reagan? That may be so, but it was investigated in depth then published in France by Renaud Foy and Antoine in 1985. On Christmas day 1984, perhaps inspired by the light that finds its way into our homes at these times of festivity, Antoine Labeyrie wrote a note that he sent to the little group at the ESO who designed the interferometer, which included himself. The note raised several questions and had an appendix entitled *Adaptive telescope with a laser probe*. The idea introduced briefly there was to inscribe an 'artificial star' in the stratosphere by means of a laser, then use this star to very quickly measure the state of the atmosphere and use the result to 'straighten out' the light wave front from a true star. Implemented by the ESO, and also on the Keck and Gemini telescopes, this idea considerably improved the performance of the adaptive optics equipping the telescopes of the VLT.

The explanation is simple. Adaptive optics has one serious limitation. To analyse the wave front that arrives at the telescope, distorted by its passage through the atmosphere, and subsequently straighten it out, those distortions must be measured very quickly, at a rate of more than a hundred times a second. This measurement can be made using the light from the star itself, provided that it is bright enough to supply enough photons in such a short lapse of time. This is not always the case, and one would then be condemned to never being able to correct for the blur of objects that were too faint, such as distant galaxies. So why not use a sufficiently bright nearby star? That's certainly a good idea, provided that it's close enough, otherwise its light will reach us through different configurations of atmospheric layers and the calculated correction will do a poor job in correcting the image of the object that interests us. The reader will recognise here the same argument as was used to motivate the interferometric dual field. Our story in Chap. 7 showed how astronomers were able to exploit the lucky circumstance of a bright object right next to SgrA*.

Did this mean that, in all other cases, and these would obviously be the most frequent, one would indeed be condemned to accept the blur in the image? Renaud and Antoine then came along with this simple remark. Suppose we have no bright enough star in the vicinity of some distant galaxy, sitting alone in a dark corner of the sky. Well, let's use a laser to *make* a star there (see Fig. 8.1). Yes, but how? It turns out that the Earth is constantly bombarded by meteorites, which produce shooting stars when they burn up in the atmosphere. This releases the sodium atoms they contain, and these build up in a layer at an altitude of about 80 km in the stratosphere. When illuminated by a laser on the Earth's surface, the atoms in a little cloud of sodium will resonate, sending back to Earth a beautiful yellow light which looks as though it is coming from an artificial stratospheric star. We thus shine the laser in the same direction as the celestial object whose image we wish to correct. The light waves from this artificial star reach the telescope after

Fig. 8.1 From the Yepun telescope, a laser shoots upwards to the stratosphere, creating there an artificial star, which is in turn used by the GRAVITY adaptive optics. The laser aims at the galactic center, where the constellation of Sagittarius shines within the Milky Way. Credit: G. Hüdepohl (atacamaphoto.com)/ESO

crossing almost exactly the same atmospheric layers and at the same time. They will have suffered the same distortions as the waves from the celestial object. All we need is a powerful enough laser attached to the telescope, to ensure that our false star is bright enough. And that's it. The blur can be corrected, even if the object is a very faint galaxy and the exposure has to last several hours. Better still, if several lasers can be mounted on the telescope with beams almost parallel, the correction can be improved even further. It is no longer limited to a few seconds of arc in the precise direction of the telescope axis, but extends over a wide field of observation, up to one minute of arc or even more.

Under Norbert's guidance and with financing by the European Community, the ESO called upon European physicists and commissioned industrial studies which eventually led to the fibre lasers equipping the VLT. These lasers comprise an optical fibre made from glass containing traces of little known rare metals with poetic names, such as ytterbium, holmium, or praseodymium. The atoms of these metals are capable of a suitable form of stimulated emission, which is the very key to the production of coherent light. Fibre lasers are powerful, stable, and reliable. Apart from a multitude of industrial applications which have derived from astronomers' war on blur, the German firm Toptica Photonics has become the global supplier of lasers for giant telescopes like Yepun in Paranal and Keck in Hawaii. Their light now inscribes artificial stars in the stratosphere.

SINFONI, MUSE, and HAWK

What is the best way to benefit from maximally sharp images, and indeed with what aims? A range of instruments are set up in fixed positions on the VLT to meet the astronomers' various needs. The first, known as SINFONI, is set up just behind Yepun's primary mirror, which has a hole at the center (a design known as Cassegrain).[2] SINFONI simultaneously analyses each of the 2048 points of an image made in the near infrared. It provides a spectral analysis of the light received by each of these points, distinguishing up to four thousand different wavelengths in its spectrum. That means that roughly eight million independent measurements are obtained simultaneously on a small celestial field, with a diameter of just 0.8 seconds of arc when the adaptive correction is excellent, and 8 seconds of arc in the least favourable case. Depending on the programme, this tiny celestial field may contain a distant galaxy whose

---

[2]https://www.eso.org/sci/facilities/paranal/instruments/sinfoni/overview.html.

composition the astronomer wishes to determine by examining its spectrum, or the motions of its hydrogen clouds using the Doppler effect.

MUSE is a magnificent instrument for observing visible light between blue and red. It is designed to observe distant galaxies, whose light reaches us from a similarly remote past, just after the first stars lit up in a Universe that was still uniformly filled with gas. Their light has crossed the Universe and been subjected to the relativistic effects of local curvature observed around black holes, or caused by dark matter. Spectral analysis of this light contains a considerable amount of information, but it requires a great deal of observation time, because these remote galaxies are so very faint. As mentioned, MUSE operates with visible light, analysing a tiny celestial field (from one minute of arc down to one tenth of that), which may contain several dozen galaxies. The telescope Yepun forms the image of this tiny field and corrects by adaptive optics. A complex optical setup then divides this image into 24 portions, themselves divided up into 48 'mini-slits' of side 0.2 seconds of arc. The light from each of these is dispersed and the spectrum thereby produced is measured by a CCD detector. The sharpness of the image, not only along the telescope axis but across the whole field, is obtained by using the four lasers and Yepun's adaptive secondary mirror.

And finally, HAWK! A most beautiful bird of prey with its all-seeing eyes, two or three times sharper than human eyes.[3] HAWK works in the near infrared and is cooled right down. It combines the maximal sharpness obtained by Yepun with an extreme sensitivity. It can measure the faintest objects, the remotest galaxies, with a combination of detectors providing 64 million pixels in each image. Moreover, it is one of the VLT instruments that 'wastes' the fewest photons on the light path—fewer than three in ten. HAWK is the distant descendant of devices conceived by Alan Moorwood, the phlegmatic Briton whose deadpan humour I mentioned when we were designing the VLT. In the past, Alan had been so busy developing image detectors that he had serious doubts about our attempts to remove the blur. He felt that they didn't get to the heart of the matter and we had long discussions about that. If only he could see what HAWK can achieve today! Thanks to the four lasers, the effects of the turbulence agitating the lowest layers of the atmosphere are almost completely removed at every point of the wide field of the image.[4]

---

[3] http://dp.mariottini.free.fr/carnets/dubai/rapaces/visionrapaces.htm.

[4] One of the many existing versions of adaptive optics in 2019 is ground layer adaptive optics (GLAO). The name refers to the fact that, in Paranal, most of the turbulence responsible for degrading our images occurs in a layer of the atmosphere about a hundred meters thick, just above ground level. This is due to the local wind systems. The proximity of the turbulence means that light rays coming from objects separated on the sky by a few tens of minutes of arc, or even more, will cross the same layer of the atmosphere and thus

Hence, in 2019, exactly 40 years after the conference in Geneva on the telescopes of the future and the discreet remark made by Gilbert Bourdet before a stony-faced Horace Babcock, adaptive optics is now generally accepted. It has become the norm at the VLT and elsewhere, equipping all the large Earth-based optical telescopes.

Interferometry in Concert

In the reddening glow of the setting sun as it dives toward the Pacific, the visitor to Cerro Paranal cannot miss the four giants on the mountaintop. With their majestic 28 m silhouettes, they completely dominate the four egg-cup auxiliary telescopes moving here and there on their rails across the immense platform that overlooks the aridity of the surrounding desert. Any interferometry between them can only be seen underground. There, carved out of the bedrock, is a huge room, completely closed, with a controlled atmosphere, housing the six delay lines which I have called the metro in the above; and then another large room which is more accessible, containing the various instruments for measuring the fringes, such as GRAVITY and PIONIER. On the floor above is the vast control room occupied by astronomers and their assistants during viewing sessions.

In the spring of 2005, those involved with the VLTI at the ESO in Garching organised a workshop entitled *The Power of Optical/IR Interferometry* (Richichi et al. 2005). So many European astronomers attended that the room the ESO scheduled for the workshop was too small. Since the aim was to get ideas for the future of the VLTI, the participants discussed new instrumentation. Apart from GRAVITY, a second instrument for measuring fringes, called MATISSE,[5] was approved. It received its first light in 2018.[6] Both couple four telescopes simultaneously in pairs, giving six different baselines, whether by combination of the giants or the egg-cup auxiliaries. MATISSE measures the fringes in the mid-infrared, the wavelength range most strongly emitted by colder matter. Another instrument was PIONIER, the prototype instrument using optical fibres, which is now open to all, thanks to its remarkable performance.

---

be simultaneously straightened out using the same reference source or artificial star. The deblurred field of observation is thus significantly extended.

[5] The person in charge of its construction, who also coordinated the various contributing research groups, was Bruno Lopez, who originally got involved with interferometry in Antoine Labeyrie's group in Nice, then worked on MIDI, one of the first instruments for measuring fringes at the VLT.

[6] https://www.eso.org/sci/facilities/develop/instruments/matisse.html.

Orchestrating interferometric matters from Garching since 2017, Antoine Mérand conducts the symphony played out by these instruments. With an acuity never before obtained at these wavelengths, they explore exoplanetary transition disks, accretion disks where matter swirls around black holes, and the atmospheres of cooling stars that fill up with dust. Images with resolutions of a few milliseconds of arc, a thousand times better than the seeing limit, are now accessible to all astronomers, even those who do not master the subtleties of fringes and their analysis. Antoine is there to help them. He belongs to that generation of young French astronomers who, at the beginning of the 2000s, entered the business of interferometry at a time when its prospects were already beginning to take shape. With Steve, he learnt how to do interferometry on the CHARA interferometer, spending a hundred or so nights a year alone among the pine trees of Mount Wilson, in winter as in summer. Then, joining Jean-Philippe Berger in Paranal, he was involved in the gradual implementation of the VLTI, before taking his place at the head of the orchestra (Mérand 2018).

Depending on the research programme proposed by a given group of astronomers and selected by the ESO after stiff competition, the group will be allocated a certain number of viewing nights. Then, depending on the sensitivity requirements, the programme will use either the four 8.2 m telescopes or the four auxiliaries, the latter being displaced on demand over the thirty positions they can occupy on the platform. The measurements obtained remain the exclusive property of the proposing group, who will skim off the key results and quickly publish the most important among them. A year after obtaining the data, the group must deposit them in a publicly accessible archive, because these research results, obtained by spending public money, must be made available to all.[7] The archives are invaluable. By consulting old measurements, we sometimes identify something that escaped earlier analysis. For instance, there may be an exoplanet, like the fifth planet orbiting the star HR8799, found by Christian Marois. Moreover, just after the end of the observation, the nightly logbook must also be made public. With the VLT, there is no way of secretly observing an object in the sky, suspected perhaps of concealing some great discovery.

Having obtained some much coveted observation time, the astronomer who makes the journey to Chile takes control of the giants' or the auxiliaries' interferometer mode in the large control room in Paranal.[8] Thanks to the collecting area of their mirrors, the giants provide exceptional sensitivity.

---

[7] https://www.eso.org/sci/observing/policies/Cou996-rev.pdf.

[8] These long-haul journeys from Europe, contributing as they do to the accumulation of CO_2 in the atmosphere and the threat of climate change, are becoming less and less necessary now with all the

However, there is a great deal of competition for viewing time and they are not allocated solely for interferometry. During nights close to the full Moon, there is less demand for other activities, and interferometry suffers less from the luminosity of the sky than observation of remote galaxies or faint stars. The egg-cup auxiliaries are less sensitive, but more often available, and above all they provide a range of different baselines because they are mobile. But whether one chooses the former or the latter, when the interferometer setup is selected, the automatic systems begin to run, the four telescopes turn obediently to point toward the chosen object in the sky, and the six sets of fringes appear on the screen.

So this is the large scale implementation of the brilliant remark made in passing by Hippolyte Fizeau a hundred and 50 years earlier! It seems so natural now. But in realising this, we should not forget the huge effort made by so many women and men who worked to create such a perfect machine. In its way, a great orchestra can also give us this feeling of absolute plenitude.

What Future for Blur?

In my story, I have sometimes used terms of war, but while the victory over the atmospheric seeing effect is real enough, some subtle refinements are still possible and will necessitate a few further battles. Adaptive optics can be extended toward visible light, increase its capacity to make out a planet a million times less bright than its star, image remote galaxies where no neighbouring star is available to analyse the wave front, and exploit the spectra of objects to distinguish more clearly between them. Interferometry, now operating with the giants as well as the auxiliaries, can still further increase its number of baselines, improve the sensitivity of its dual field, and better use the all too scarce photons. The prospects for the 10 years to come are already largely determined for both alike.[9]

However, on the ground as in space, the quest for the ultimate detail will still come up against the diffraction barrier. From the beginning of this story, we have known that this unavoidable limit is due to the very nature of light

---

improvements in Internet communications. It is often possible today to carry out observations at the VLT from a research center in Europe.

[9]In the spring of 2019, the ESO organised a workshop, similar to those held over previous decades, in order to plan for Paranal in 2030, when the E-ELT will be up and running. Early in 2020, a proposed upgrade of GRAVITY, named GravityPlus, was given top priority by the ESO's Scientific and Technical Committee.

itself.[10] The blur generated by diffraction depends only on two parameters: one is the size of the telescope mirror, or the maximal length of the interferometric baseline when several telescopes are coupled, and the other is the wavelength of the light. The larger the former, the shorter the latter, the better the resolution will be. These are ways forward to continue the campaign against blur.

It is possible to isolate ever smaller details in an image by reducing diffraction. The only thing is that the reduced volume must emit enough light energy to be detected. This means that, given the diameter of mirror used to pick up the light, it must be measurable within a reasonable time of exposure, i.e., seconds, minutes, or even hours, assuming that the detector doesn't waste any photons and there is no background noise.[11] The possibility of further increasing the mirror area does not mean there will be no limit, because at some point the resolved detail will not be bright enough to remain measurable without excessive exposure times. The impact of this limitation must be studied in any project for a new super-resolving instrument. What good would it be to improve the diffraction limit if the resulting signals become too faint?

The explorations of exoplanets and black holes await answers from new supergiant telescopes and new interferometers on the ground or in space, and these will in turn lead to new questions and problems.

Cerro Armazones

From the top of Cerro Paranal, one can see to the east across the Atacama desert as far as the peaks of the Andes. In the daytime, the air is so transparent that the summit of Cerro Armazones, 23 kilometers away, seems deceptively close. A long winding road snakes its way up to it and the traffic on it was heavy in 2019. Indeed, the fourteen member countries of the ESO decided in 2011 to build a 39 m optical telescope at the top of this mountain. First light for the European Extremely Large Telescope (E-ELT) is scheduled for 2025 (see Fig. 8.2). It will relegate the VLT giants to second place. Its scientific programme[12] will give

---

[10]With the dual field method, we discussed the possibility of differential measurements, which do not produce images in the normal sense. Their astrometric accuracy in measuring the separation between two objects can go well beyond the diffraction limit, but there is no contradiction here with this limit affecting the resolution of an image. As noted in Chap. 7, physics can provide many examples of extremely precise differential measurements, such as those used to detect gravitational waves.

[11]These ideal conditions are hard to achieve on Earth, where there are many sources of background noise, even though our photoelectric visible and infrared light detectors (CCDs or equivalent) are today almost perfect.

[12]https://www.eso.org/sci/facilities/eelt/.

Fig. 8.2 An artist's rendering of the 39-m European Extremely Large Telescope, being built on Cerro Armazones, close to Cerro Paranal in the Atacama Desert. Four laser beacons are creating artificial stars for the adaptive optics. In May 2017, the first stone was laid in the presence of Michele Bachelet, President of Chile, and the telescope will open in 2025. Credit: ESO/L. Calçada

pride of place to black holes and exoplanets, working with visible and infrared wavelengths like the VLT, but astrophysics is not something that can be divided up into strictly separate fields with no leakage between them. Knowledge of the galaxies and the stars, the distant past and more recent times, the largest scale and the finest detail, everything is necessary because everything is related in the Universe.

The adaptive optics of the E-ELT on which Norbert and his team are working will be more complex than the system on Yepun. To deblur the images, the adaptive and deformable mirror will be 2.6 m in diameter, more than twice the linear dimension of the one on Yepun. This mirror is no longer the secondary mirror of the telescope, but it is placed in fourth position in the optical train which forms the focal image. It will comprise 7000 actuators to deform its surface. Instead of four laser beams, there will be six to create artificial stars, arranged in a circle within the field of view.[13] The camera MICADO will be one of the so-called first light instruments, since it will be among the chosen few to receive first light in 2024. The contract for the camera was signed in 2015 between Reinhard Genzel, in charge of realising it, and the Dutch astronomer Tim de Zeeuw, then Director General of the ESO.

The camera will receive near infrared light whose distorted wave fronts will have been straightened out twice, once by the 2.4 m deformable mirror and a second time in more detail by a second mirror actually inside the instrument— our friends Yann Clénet, Gérard Rousset, and Eric Gendron are in charge of this. The target resolution for its images can be illustrated by the angular size of 1.5 milliseconds of arc associated with the smallest pixels. The target for astrometric measurements will be thirty times better, with a value of 50 microseconds of arc. The MICADO consortium is a partnership between French institutes in Meudon and Grenoble and German ones in Garching and Heidelberg, already working closely together on the VLT, with others in Italy, the Netherlands, and Austria. The programme announced for MICADO is fascinating. No surprise to find the black hole at the galactic center on the list, but it's not the only black hole to appear there. Another target is the intermediate mass black hole, at ten thousand times the mass of the Sun, which seems to be lurking within the globular cluster Omega Centauri. This beautiful cluster of stars in the southern hemisphere is visible with the naked eye and its apparent diameter viewed from the Earth is 30 min of arc, the same as the Moon's.

Across the Atlantic, another new generation telescope is being prepared for some time around 2027. It is called the Thirty Meter Telescope, this referring to the size of its mirror, made up of a tiling of hexagonal segments, like the mirror of the E-ELT. It is a distant offshoot of the ideas that gave rise to the two Keck telescopes. In Chile, the Giant Magellan Telescope (GMT), built by the United States and a large international collaboration, will combine seven 8.4 m telescopes on the same mount. Here we recognise a descendant of the Large Binocular Telescope on Mount Graham in Arizona along with the ideas of Roger Angel, whose 'mirror factory' in Tucson will polish these seven mirrors.

There will soon be adaptive optics everywhere. Perhaps it would not be going too far to say that these huge Earth-based telescopes with their glorious future would not have seen the light of day if adaptive optics had not shown that the seeing ("those many trembling Points confusedly and insensibly mixed with one another", as Newton described it) could be overcome. Otherwise, the

---

[13] https://en.wikipedia.org/wiki/Extremely_Large_Telescope.

drive to obtain fine detail in astronomical images would have led astronomers to favour the construction of large spaceborne telescopes. On the other hand, there will be at least one, the successor of the Hubble Space Telescope. In 2021, NASA will launch the James Webb Space Telescope into space. It will have a mirror with diameter 6.5 m.[14] Despite this diameter and the huge technological challenge of assembling the mirror in orbit, the resolution in JWST images will be six times less than in the images produced by MICADO on the E-ELT for the same wavelength. The latter will only be exceeded by the VLTI and the interferometers that supersede it. And the European Space Agency, partner[15] in the JWST, has chosen not to pursue plans for a comparable large telescope in space.

Habitable Worlds and Future Optical Interferometry

There is a word I have only used once so far, although it is abundantly present in research publications on exoplanets, and also in the media. This is the word 'habitable' (Schneider 2018). It is very likely one of the words that will feature prominently in this research over the coming decades, representing as it does a key issue in the search for exoplanets (Seager 2013).[16] And in fact, it was first used many years ago, in 1959, by Su-Shu Huang, a scientist at the University of California in Berkeley, in a paper with the title *The problem of life in the Universe and the mode of star formation* (Huang 1959). The main part of the discussion centered around a comparison between the time required for the evolution of stars, already well understood, and the time scale for the evolution of life. Today, the discussion has moved on somewhat, since the first problem is to see whether the conditions on such and such a planet lend themselves to the existence of some form of life like the one we know here on Earth.

Given the very reliable assumption that such life forms could only appear and perdure in the presence of liquid water, habitability would be defined by examining the physical and chemical conditions, like temperature, pressure, and chemical composition, which would actually allow the presence of liquid water. The concept can be refined, but simplifying somewhat, and noting that the water molecule (H_2O) is abundant in the Galaxy, the whole problem

---

[14] https://www.jwst.nasa.gov.

[15] Among the various instruments at the focal point of the JWST will be MIRI, which includes a coronagraph, proposed by Daniel Rouan. The Paris Observatory in Meudon and the French atomic energy authority, the CEA in Saclay, are in charge of that. In this NASA mission, French teams maintain their presence.

[16] See also https://en.wikipedia.org/wiki/Circumstellar_habitable_zone.

reduces almost to the question of how far the planet is from its host star—not too close because that could evaporate the oceans (100 °C), nor too far away because ice could then cover the whole surface (0 °C). The presence of an atmosphere would be relevant here, since it is an atmospheric greenhouse effect that helps to keep the Earth's surface above 0 °C. Naturally, one might criticise the habitability criterion for being rather human-centered, or at least for limiting considerations to just one life form. This is not the place to go into this lively and interesting debate (Lequeux and Encrenaz 2019), but note that there are already sixteen exoplanets known to be situated in the habitable zone of their host star. Proxima Centauri b is one of these and has been closely scrutinised by SPHERE, as discussed above.

The question of life in the Universe is a huge subject which goes beyond our scope here. However, we may be certain that during the present century, the study of exoplanets will supply the main details.[17] The capacity of future instruments to contribute to this investigation would thus appear to be one of the most important things to consider in their design. Their planned angular resolution and sensitivity should be assessed with respect to these objectives.

For this reason, given that Europe has chosen to invest materially and financially in the E-ELT, the VLT interferometer will remain in use beyond 2030, all the more so in that the work load placed on the Paranal giants is likely to decrease when the new giant is up and running. Many things remain possible to improve the performance of the VLTI, and many new ideas will arise to get round the limitations due to the Earth's atmosphere, like the idea of the dual field method using optical fibres 20 years ago, or the idea of the artificial laser star. There would be nothing to stop building two more egg-cup auxiliaries, either fixed or mobile, increasing the number of possible baselines to fifteen[18] and hence using aperture synthesis to improve the resemblance between the images and the target object.

In the United States, our friends involved in interferometry have not given up either, since the CHARA array on Mount Wilson has confirmed, if confirmation were needed, that interferometry is indeed a frutiful approach. The observation and study of exoplanets have brought together an ever increasing community of astronomers, analysing the thousands of transits accumulated by NASA's Kepler mission. A new project therefore came into being in 2013, called Planet Formation Imager (PFI).[19] Like the ALMA radiotelescope

---

[17]Media interest cannot be ignored since this is a subject that sells extremely well. However, we should be cautious of overselling. That could be counterproductive in the long run, because no one is in a position to say what will become of this conjecture!

[18]Recall that the number of baselines for six telescopes which can all be combined pairwise is $6 \times 5/2 = 15$.

[19]http://planetformationimager.org/.

array, this project will doubtless be carried out by a worldwide partnership, organised around John Monnier, an astronomer as able as he is discreet, who values and admires the links he has with Europe. Indeed, the project brings together a good many of those encountered in the present story, including Jean-Philippe Berger, Sylvestre Lacour, Stephen Ridgway, and others, like the Belgian astronomer Jean Surdej, whom we could also have mentioned. As the name makes clear, this Earth-based interferometer is designed to obtain still sharper images of the disks and exoplanets forming in them. The aim will be to better understand exoplanetary formation systems and the migrations they produce, by pushing back the limits imposed by diffraction. In 2019, the definition of the Planet Formation Imager looks a bit like that of the VLTI back in 1982. For example, the outline considers ten to twenty telescopes of diameter 4 m (gain in sensitivity) or more, working in adaptive mode in the near infrared. Its baselines could be as long as ten kilometers, giving a resolution as good as 500 microseconds of arc, a hundred times better than the VLTI.[20] The light beams are carried in vacuum to the point of fringe formation, using hollow optical fibres, and a lot of care is taken over the scarcity of the photons. Another possibility is to use the method adopted by Charlie Townes, at least partially.[21] Many creative ideas may still crop up and the imaginative phase remains wide open (Monnier et al. 2014, 2018). And Antoine Labeyrie is still actively involved, exploring the idea of a hypertelescope of his own design.

In the present account, we have often referred to the complementarity between interferometry in the infrared and visible, on the one hand, and at millimeter and submillimeter wavelengths (on the Plateau de Bure, ALMA, and the Event Horizon Telescope), on the other. The successors of these millimeter arrays are not yet in the design phase, but they are sure to come and this complementarity will continue.

In Space

Just after qualifying to conduct research, the young Sylvestre Lacour used GRAVITY to determine the astrometry of an exoplanet in the first spectacular

---

[20]Resolution without sensitivity is useless if the target objects are not bright enough, as we saw above. Any instrument project must be clear about this point. Likewise, a large number of telescopes sharing the light in order to operate simultaneously ends up reducing the sensitivity, as we saw for OVLA.

[21]As discussed in Chap. 4, with the Infrared Spatial Interferometer (ISI) on Mount Wilson, Townes and his team in Berkeley used a method analogous to the one applied in radiointerferometry, as regards the light detection technique. The fringes are formed at a very precisely specified wavelength, the reference being supplied by a laser. The measurement is then easier, but it limits the sensitivity of the instrument.

observation discussed above. Some time before, within our group at the Paris Observatory in Meudon,[22] he became fascinated by Beta Pic b and suggested observing its transit in front of the host star. The disk in this case is viewed edge-on and contains the orbit of the planet. Since the orbit is known, the periodic transits can be predicted. One of these was announced for 2018. So why not try to observe it from space using a nanosatellite? This novel concept has been made possible by the ever improved miniaturisation of electronics, computer systems, and optics. Costs can be considerably reduced because several dozen nanosatellites can be launched by a single rocket, each with its own specific mission. Sylvestre built the satellite, with a mass of 3.9 kg, and quite naturally called it PicSat, giving it all the stability and sensitivity that would be needed to spot the transit, monitor it, and transmit the measurements to the Earth. The light was to be carried in a single-mode optical fibre—Sylvestre's office at the observatory was next door to Vincent Coudé du Foresto's! But unfortunately, 3 months after its launch in January 2018, the little PicSat stopped returning data.

NASA's highly successful Kepler mission ended in 2018. Using the trick of spotting transits, it clearly demonstrated the unexpectedly high frequency of exoplanetary systems. It was followed by the Transiting Exoplanet Survey Satellite, launched in 2018, which also looks for transits in front of stars within a range of 600 light-years from Earth. However, neither of these can provide direct images of exoplanets. It certainly seems likely that we will one day see new spaceborne missions implementing the ideas of Mike Shao with the Space Interferometry Mission or the Terrestrial Planet Finder, or again those of Antoine Labeyrie with his TRIO array flying together in space, or other ideas that inspired Alain Léger, Jean-Marie Mariotti, Daniel Rouan, and Jean Schneider in the Darwin project![23] If we aim for baselines of a kilometer or more in the hope of at last imaging the surface of an exoplanet—even with just a few pixels—, the sensitivity needed will impose quite a large mirror diameter, of the order of a meter at least, as noted above. These mirrors must fly independently and in formation, with extremely precise relative positions, since the aim is to do interferometry. Their separations must be measured by laser and continually stabilised to within a small fraction of the wavelength of the light they are collecting, i.e., a few nanometers. Fortunately, the dynamical

---

[22]The research center in Meudon has often been mentioned in this story. Although the Paris Observatory has seven different research centers called 'departments', the one I have been mentioning, to which I still belong, is the *Laboratoire d'Etudes Spatiales et d'Instrumentation en Astrophysique* (LESIA), which is the largest.

[23]These ideas are still alive and well: Defrère et al. (2018).

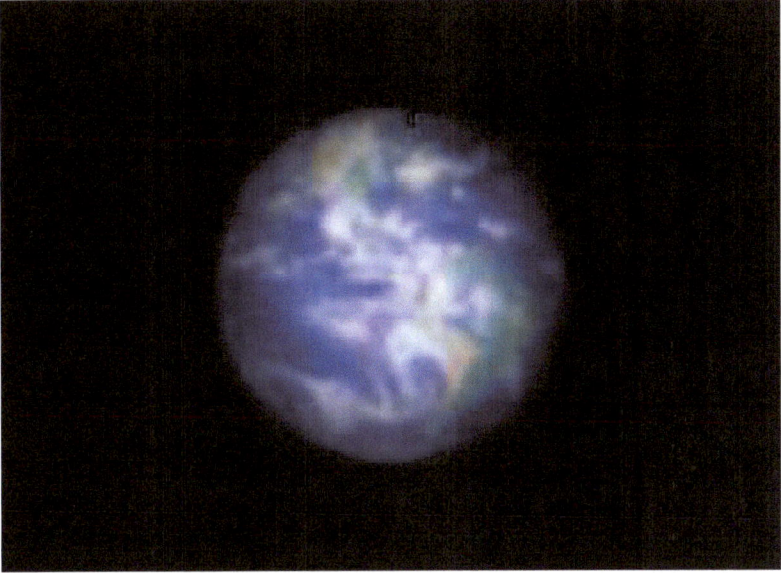

Fig. 8.3 Dreaming of the future. During the 1980s, Antoine Labeyrie dreamt of an optical space interferometer with kilometric baselines. He simulated the image of an exoplanet identical to Earth, as seen at 30 light-years from us if seasonal vegetation would look like the green Amazonian area shown here. Indeed, the residual image blur is the result of the limited resolution. Image courtesy of Antoine Labeyrie

perturbations will be very weak and slow in space, so the idea is not completely inconceivable.

The discovery of gravitational waves makes it very likely that a European mission will be sent into space. This mission, called LISA, was first mooted at the beginning of the 2000s.[24] A possible date would be 2034. The mission would require extreme navigation accuracy for the satellites, arranged millions of kilometers apart. An exploratory mission called LISA Pathfinder was launched by the European Space Agency in 2015 to test the feasibility of maintaining this level of accuracy in the positioning of such satellites. The conclusion was affirmative.[25]

So these technologies explore methods that will one day be invaluable for setting up a kilometer-baseline interferometer in space to study the details of exoplanets, the close environments of black holes, or indeed other target objects. With such an instrument operating in visible light, the image of an

---

[24] https://www.elisascience.org/articles/lisa-mission/lisamission-gravitational-universe.
[25] https://en.wikipedia.org/wiki/LISA_Pathfinder.

exoplanet would no longer be a dimensionless geometric point. The diffraction limit achieved by the interferometer would reveal the details of the planetary surface on a handful of pixels, including maybe clouds, forests, or other, unknown landscapes.

And as there will always be young researchers ready to join the endless quest for more detail in our view of the Universe, contributing their most creative years with the pleasure of belonging to this international community, the story of the blurred image will not end here.

9

Epilogue

> For after all what is man in nature? A nothing in relation to infinity, all in relation
> to nothing, a central point between nothing and all [...]—Blaise Pascal, *Pensées*

As a child, shortly after the Second World War, I was sent to spend a summer on the German banks of the Rhine. Language-learning trips were not yet the fashion. The aim was above all to prepare young people who would actively seek to make a new Europe, after three wars between France and Germany in less than a century. I recall the shock I felt when I crossed the town of Cologne, still in ruins, with only the cathedral still standing. A few years later, our year at the *École normale supérieure* was invited to visit the European Coal and Steel Community in Luxembourg. It was the first step toward the future European Union. And so I discovered the idea of Europe.

With the creation of the *Centre européen de recherches nucléaires* (CERN) in 1951, then the European Southern Observatory (ESO), the European Space Agency (ESA), and a few other research institutions, and with the European Research Council to attribute the funds provided by the Union, European scientists have shown the tremendous potential of their intellectual, technological, industrial, and financial collaboration, and the extraordinary things that can be achieved this way, working in their respective universities to maintain a unique scientific tradition. As stated in CERN's founding charter, "science is placed at the service of peace". Let's hope this remains so.

Over the last half century, the campaign against blur was not exclusively European. Of course, no scientific undertaking can be truly successful if it does not accept the universality of the challenges and the universality of

© Springer Nature Switzerland AG 2020
P. Léna, *Astronomy's Quest for Sharp Images*, Astronomers' Universe,
https://doi.org/10.1007/978-3-030-55811-6_9

exchange. But without ignoring this cross-fertilisation, the story told here will have shown that our European collaboration took us among the highest levels in the world, and sometimes to the very highest. From the shores of the Mediterranean to the Arctic frosts, everyone had their part to play, depending on their own story and their own style. One of my aims was to highlight the often decisive role played by cooperation across the Rhine and the links it forged. I hope I have also shown how much I personally, and we all, owed to the transatlantic cooperation with our friends in the United States. Likewise for Chile, among the emerging nations, which was in a difficult situation in those days but is nowadays well-placed in world astronomy.

In my introduction, I explained how much the fight against blur owes to the energy and creativity of young people who have chosen to go into research. They are afraid of nothing, neither the personal risk of venturing into unknown territory with uncertain gain, nor the lack of self-confidence which is often the prerogative of the best minds, nor the material hardships of unreliable job situations, nor the consequences of being posted abroad which can lead to family problems, nor even the inevitable failures. Their enthusiasm would always provide the universal remedy to such risks. These astronomers, men and women, have beaten the blur, and their reward is a cornucopia of discoveries and new questions. This is what I have tried to show and to share. But I have also written this for the generation to follow, those still in high school, university, or engineering school. In a world which sometimes seems to believe only in accumulating financial wealth and in which words often stand alone in the place of reality, the demands of research are sometimes seen as the concerns of another age, little understood by the powers that be. It is my view that young people deserve another message.

At this point, I cannot help but reflect upon the words of Blaise Pascal, when he writes of the disproportion of man.[1] This story has taken us toward the multitudes of exoplanets, these worlds where radically new and at first sight improbable things are likely to emerge. It may be that the whole range of the possible, everything authorised by the laws of Nature, will be found there, providing an inexhaustible source of amazement. As Pascal wrote:

> [...] let imagination pass beyond, and it will sooner exhaust the power of thinking than Nature that of giving scope for thought.

---

[1] Pascal (1954). See also https://www.gutenberg.org/files/46921/46921-h/46921-h.htm for an English version.

Then, turning our gaze toward Sagittarius, we came upon a prodigious concentration of matter which is no doubt a black hole. Nature has placed it within reach of our telescopes and our images. We have even brushed right up against its event horizon, the point at which any state yet conceivable simply disappears.

Metaphor or reality, this story has brought us life in the novel splendour of the dawn, death in the disturbing annihilation of the dusk. A central point between nothing and all, this is where the human being stands on this Earth. A being who stands in wonder, who discovers and thinks, who loves and suffers, a reed swaying in the wind, perhaps, but grappling with its future.

A

Distances and Angles

Table A.1 Angles characterising blur. One minute of arc (arcmin) is equal to 60 seconds of arc (arcsec)

Configuration	Achievable resolution angle
Resolving power of the human eye (20/20)	Approx. 1 arcmin = 1/60 degree
Seeing limit	0.5–1.5 arcsec (average values in Paranal)
Diffraction limit of MACAO adaptive optics (near infrared, 8.2 m mirror)	50 milliseconds of arc
Interferometry diffraction limits (near infrared)	
Four VLT giants combined (130 m baseline)	3 milliseconds of arc
Four VLT auxiliaries combined (200 m baseline)	2 milliseconds of arc
Differential astrometry (GRAVITY, dual field)	Approx. 10 microseconds of arc
ALMA interferometer (wavelength 1 mm, baseline 16 km)	12 milliseconds of arc
VLBI interferometric diffraction limit (maximal terrestrial baseline)	
Wavelength 10 cm	2 milliseconds of arc
Wavelength 1 mm	20 microseconds of arc

© Springer Nature Switzerland AG 2020
P. Léna, *Astronomy's Quest for Sharp Images*, Astronomers' Universe,
https://doi.org/10.1007/978-3-030-55811-6

Table A.2 Distances. Note that 1 astronomical unit $\approx$ 8 light-minutes $\approx$ 150 million km is the distance from the Earth to the Sun, and 1 parsec $= 3 \times 10^{16}$ m $= 206,000$ AU. The Schwarzschild radius of SgrA* is given by the formula $R_S = 2GM/c^2$, where G is Newton's gravitational constant, c is the speed of light, and M is the mass of the black hole, in this case approximately 4 million times the mass of the Sun

Object	Distance [m]	Distance [light-time]	Distance [AU]
From the Sun to Jupiter	7.8×10^{11} m	42 light-minutes	5.2 AU
From the Sun to Proxima Centauri	4×10^{16} m	4.2 light-years	267,000 AU
From the Sun to Beta Pictoris	–	63 light-years	4×10^6 AU
From the Sun to SgrA*	2.44×10^{20} m $= 8000$ parsec	26,500 light-years	1.6×10^9 AU
From the peribothron of S2 to SgrA* (distance projected on the sky)	1.5×10^{13} m	14 light-hours	100 AU
Schwarzschild radius of SgrA* (mass 4 million solar masses)	1.2×10^{10} m	40 light-seconds	0.08 AU

Table A.3 Angles as observed from the Earth. Note that the shadow of the black hole is the collection of paths of photons that did not escape, but were instead captured by the curvature of spacetime. R_S is the Schwarzschild radius

	Angle
Lunar disk	30 arcmin
Jupiter (average disk diameter)	Approx. 20 arcsec
Debris disk Beta Pic	Approx. 8 arcsec
From HR8799 to its planet e (the outermost planet)	400 milliseconds of arc
A Jupiter orbiting Beta Pic (diameter of the planet)	20 microseconds of arc
Black hole shadow	Approx. 55 microseconds of arc
Variable distance between SgrA* and S2 on its orbit	100 to 12 milliseconds of arc
Distance between the peribothron and SgrA* (100 AU or 1250R_S)	12 milliseconds of arc
Radius of hotspot orbit (3 to 5R_S)	30–50 microseconds of arc
Distance between SgrA* and the star IRS7	5.5 arcsec
Distance between SgrA* and the star IRS16C	1.23 arcsec

B

Telescopes and Instruments

Adaptive Optics Facility (AOF) Adaptive optics system for the telescope Yepun in Paranal, implemented in 2016 and comprising a deformable secondary mirror, wave front sensors, and four lasers mounted on the telescope to create artificial stars.

Adaptive Optics Near Infrared System (ADONIS) Adaptive optics system set up on the ESO's 3.60 m telescope in La Silla, Chile, in 1993, and operating in the near infrared. It was used with the cameras COMIC and SHARP2+, and was available for general use until the beginning of 2000.

Antu One of the VLT's four 8.2 m telescopes. Also called UT1. In Quechua, the name means 'Sun'.

Astronomical Multi-BEam Combiner (AMBER) First generation instrument for detection of fringes produced by the VLTI, operating in the near infrared and set up at the interferometric focal point. When it went into operation in 2007, AMBER coupled three telescopes. Built by the Côte d'Azur and Grenoble Observatories.

Atacama Large Millimeter Array (ALMA) An array of 66 radiotelescopes on Cerro Chajnantor (Chile) at an altitude of 5100 m, operating at millimeter and submillimeter wavelengths since 2011. ALMA is the result of an international partnership, including in particular Europe, the United States, Japan, and Chile.

© Springer Nature Switzerland AG 2020   **241**
P. Léna, *Astronomy's Quest for Sharp Images*, Astronomers' Universe,
https://doi.org/10.1007/978-3-030-55811-6

Auxiliary Telescopes (AT) Four mobile 1.8 m telescopes included in the infrastructure of the VLT Interferometer (VLTI) in Paranal. Referred to as 'egg-cups' in the text, due to their general appearance.

Cambridge Optical Aperture Synthetic Telescope (COAST) Optical interferometer comprising five 40 cm telescopes on baselines up to 47 m. Set up in Cambridge (UK) in 1997.

Canada–France–Hawaii Telescope (CFHT) A 3.6 m optical telescope at the top of Mauna Kea (4100 m) on the Big Island of Hawaii. Inaugurated in 1979, it was developed by a partnership between Canada, France (CNRS), and the University of Hawaii in Honolulu.

Center for High Angular Resolution Astronomy (CHARA) Interferometer set up by the Georgia State University in the United States. It comprises six 1 m telescopes operating at visible and infrared wavelengths. It is set up on Mount Wilson, near Los Angeles, and its maximal baseline is 600 m.

ComeOn The first astronomical adaptive optics system in the world, developed by CGE, the Paris Observatory in Meudon, the ESO, and ONERA, and operating in 1989. It was the result of a partnership between France and the ESO, installed in 1990 on the ESO's 3.6 m telescope in La Silla (Chile). It was superseded by an improved version called ComeOn+, then replaced in 1993 by ADONIS.

CONICA Near infrared imager and spectrograph set up in 1998 at the Cassegrain focus of the Yepun telescope in Paranal, and still in use in 2019.

Darwin European space mission for the study of exoplanets, proposed in 1996 but abandoned in 2007–2008.

European Extremely Large Telescope (E-ELT) Telescope with diameter 39 m, set up by the ESO on Cerro Armazones (Chile), not far from Paranal. First light is scheduled for 2025.

Fiber Linked Unit for Optical Recombination (FLUOR) Instrument for forming and measuring fringes in the infrared region, using single-mode optical fibres for the first time. Built in 1991 at the Paris Observatory, it was set up on the IOTA interferometer in Arizona. FLUOR considerably improved the accuracy of fringe measurements.

Gemini Two 8.1 m optical telescopes, one on Mauna Kea on Hawaii's Big Island, the other on Cerro Pachon in Chile. An initiative of the National Science Foundation (United States), they were realised by an international partnership and have been operating since 2000.

Giant Magellan Telescope (GMT) Giant optical telescope with seven 8.4 m optical mirrors, combined to form a continuous surface with an equivalent diameter of 22 m, set up in Las Campanas (Chile). Built by a consortium of American universities, with participation by Australia, Brazil and Korea. This telescope will be operational in 2025.

Giotto Space probe sent by the European Space Agency to observe the nucleus of Comet Halley and photograph it at close range in 1986.

Grand Interféromètre à deux Télescopes (GI2T) Interferometer operating in visible light, including two moveable 1.5 m telescopes. Set up in 1984 by Antoine Labeyrie on the Plateau de Calern, near Grasse (France). Ceased operations in 2004.

GRAVITY Instrument for forming and measuring fringes in the infrared region, set up at the interferometric focus of the VLT in Paranal. This instrument, equipped with a specific kind of adaptive optics and a dual field, can be supplied by all or a subset of the four 8.2 m telescopes, or the four 1.8 m mobile auxiliary telescopes. Began operations in 2016.

Hale Telescope This is a 5 m telescope on Mount Palomar (California), run by the California Institute of Technology since 1947.

HAWK High acuity near-infrared wide field imager, corrected by adaptive optics, in operation since 2007 at one of the Nasmyth focal points of the telescope Yepun in Paranal.

Hokupa'a Adaptive optics system set up on the Gemini North telescope in Hawaii in 1999. This system works by analysing the curvature of the wave front.

Hooker Telescope A 2.5 m optical telescope set up on Mount Wilson (California) in 1917. Edwin Hubble used this telescope to observe the recession speeds of galaxies, thus establishing the expansion of the Universe.

Hubble Space Telescope (HST) This 2.4 m space telescope was launched by NASA with European participation in 1990. Its cameras detect visible, near infrared, and near ultraviolet wavelengths. Scheduled to operate until 2020.

Infrared and Optical Telescope Array (IOTA) Interferometer comprising 3 mobile telescopes (siderostats), set up on Mount Hopkins (Arizona) in 1993 and operating in the near infrared with a maximal baseline of 35 m. This was built and is now run by a consortium of American universities.

InfraRed Astronomical Satellite (IRAS) The first satellite to carry a telescope observing the whole sky in the infrared region, including near, mid, and far infrared. Launched by NASA in a partnership with Great Britain and the Netherlands in 1983, IRAS operated for only 10 months, as it had a limited supply of cryogenic fluid to cool the detectors.

Infrared Space Observatory (ISO) The second spaceborne observatory to carry a telescope observing the whole sky in the infrared region, including near, mid, and far infrared. Launched by the European Space Agency in 1995, it operated until 1998.

Infrared Spatial Interferometer (ISI) This interferometer was developed from 1974 by Charles Townes' research group in Berkeley (California) and works on the principle of heterodyne light detection at a wavelength of 10.6 μm, with three mobile 1.65 m telescopes installed on Mount Wilson (California). The maximal baseline is 32 m.

Institut de RadioAstronomie Millimétrique (IRAM) Founded in 1979, this institute was the result of a partnership between France (CNRS), Germany (Max Planck Society), and Spain (Instituto Geografico Nacional). There are two sites, one with a millimetric radio interferometer (Noema) on the Plateau de Bure in the Dévoluy (France) and a 30 m radiotelescope at Pico Veleta near Grenada (Spain).

Interféromètre à 2 Télescopes (I2T) These two optical telescopes each have a diameter of 25 cm, and stand on a 12 m baseline. It was the first interferometer built by Antoine Labeyrie, set up in Nice, then on the Plateau de Calern near Grasse in the south of France in 1974. It was superseded by GI2T 10 years later.

James Webb Space Telescope (JWST) This 6.5 m telescope will observe visible and (near and mid) infrared wavelengths and is intended to supersede the Hubble Space Telescope. Launch is scheduled for 2021. This project is led by NASA in a partnership with the European Space Agency and the Canadian Space Agency.

Keck Telescope Two 10 m telescopes operating in the visible and (near and mid) infrared, set up at the top of Mauna Kea on Hawaii's Big Island. The possibility of interferometric operation, including also two auxiliary telescopes, has been shelved since 2012. This instrument belongs to the California Institute of Technology and was financed by the W.M. Keck Foundation and other support funds.

Kueyen One of the four giant telescopes of the VLT. Also called UT2, the name means 'Moon' in Quechua.

Large Binocular Telescope (LBT) Two 8.4 m telescopes are supported on the same mount and can operate independently or interferometrically. The LBT began operations in 2006. The two telescopes observe in the visible and (near and mid) infrared, from the top of Mount Graham (Arizona), at an altitude of 3270 m. They were built by an international partnership, led by the University of Arizona.

Laser Interferometer Gravitational Wave Observatory (LIGO) This instrument is designed to detect gravitational waves and includes two interferometers, one in Washington State and the other in Louisiana. It has been running since 2015, in partnership with the French–Italian gravitational wave detector VIRGO.

Laser Interferometer Space Antenna (LISA) Project to build a gravitational wave detector in space, still under study at the European Space Agency (2019). LISA would comprise three independent but connected satellites, located at the corners of triangle with the dimensions of the Earth's orbit around the Sun.

McMath–Pierce Solar Telescope A set of three telescopes, supplied by three moveable plane mirrors (heliostats), including one 1.6 m telescope, designed to observe the Sun at visible and near infrared wavelengths. Set up by the National Science Foundation at Kitt Peak (Arizona). Built in 1962 and equipped with adaptive optics.

Mark I–III Stellar Interferometer A series of three optical interferometers set up on Mount Wilson (California) by Michael Shao and Mark Colavita at the California Institute of Technology, with a maximal baseline of 31 m.

Mayall Telescope This is a 4 m optical telescope at Kitt Peak (Arizona), run by the National Optical Astronomy Observatory (National Science Foundation) since 1973.

Melipal One of the four 8.2 m telescopes of the VLT. Also called UT3, the name refers to the constellation of the Southern Cross in Quechua.

Mid-Infrared Interferometric Instrument (MIDI) First generation fringe detection instrument of the VLTI, operating in the mid-infrared. Installed in 2002, MIDI receives the beams from two 8.2 m telescopes to produce fringes. Built by a consortium of research centres in Germany, France, and the Netherlands. Replaced by MATISSE in 2018.

Multi-Adaptive Optics Imaging Camera for Deep Observations (MICADO) Camera to be placed at the Nasmyth foci of the future E-ELT, including advanced adaptive optics. Scheduled for operation in 2025, it is being built by a consortium of European research groups.

Multi-AperTure mid-Infrared SpectroScopic Experiment (MATISSE) Fringe detection instrument of the VLTI in Paranal, operating in the mid-infrared since 2018.

Multi-Application Curvature Adaptive Optics (MACAO) Adaptive optics system based on wavefront curvature analysis with 60 active elements, set up in 2003 at each of the fixed (Coudé) foci of the four 8.2 m telescopes of the VLT. These systems correct the beams supplying the combined interferometric focus.

Multi-Unit Spectroscopic Explorer (MUSE) Spectrograph able to observe many objects in the field of view simultaneously, in visible light. Set up at the Nasmyth focus of the telescope Yepun in 2014, with AOF adaptive optics. Under the responsibility of the Lyon Observatory (France).

NAOS–CONICA (NaCo) Ensemble combining the NAOS adaptive optics system with the CONICA camera. Used on the VLT telescopes Yepun, then Antu, since 1998.

Nasmyth Adaptive Optics System (NAOS) First adaptive optics system set up on Yepun at the VLT in 2001. Its deformable mirror has 185 actuators. Built by a French consortium (Grenoble, Meudon, and ONERA).

National New Technology Telescope (NNTT) Project to build a very large optical telescope, with diameter 16 m, studied by the United States National Science Foundation in the 1980s. Never completed in its original form (see Gemini).

New Adaptive Optics Module for Interferometry (NAOMI) Adaptive optics instrument equipping each of the auxiliary telescopes (AT) of the VLTI, before the beams are sent into the interferometry laboratory to form fringes. Built by the Grenoble Observatory (France) and installed in 2018.

New Technology Telescope (NTT) This is a 3.58 m optical telescope, built by the ESO and set up in La Silla (Chile) in 1982. The NTT was the first telescope in which the shape of the primary mirror could be permanently adjusted (active optics).

Optical Hawaian Array for Nanoradian Astronomy (OHANA) The name was inspired by the Hawaiian word 'O'hana', meaning 'family'. This was a project to couple the large telescopes on Mauna Kea (Hawaii), in particular, the CFHT, the Keck, Gemini North, and Subaru, to form a near infrared interferometer. It was led by the Paris Observatory between 2000 and 2008, but was only partially successful.

Optical Very Large Array (OVLA) Idea for an optical interferometer with 27 mobile 1.5 m telescopes, studied by Antoine Labeyrie over the period 1984 to 2000.

Palomar Testbed Interferometer (PTI) Small interferometer set up by Cal-Tech on Mount Palomar in 1995, designed to test the ideas subsequently implemented on the Keck Telescope interferometer. It operated between 1995 and 2008. The dual field idea was developed here.

Phase-Referenced Imaging and Microarcsecond Astrometry (PRIMA) Project involving various devices to be able to use the VLTI for extremely accurate astrometry. It included in particular laser metrology and the possibility of dual field, i.e., simultaneously obtaining fringes for two very close celestial objects. The project, launched by the ESO in 1996, was never set up on the ATs, but the studies were important for the implementation of the GRAVITY instrument.

Planet Formation Imager (PFI) Earth-based optical interferometer project for the study of exoplanets and their formation, combining several large telescopes on kilometric baselines. A study was launched in 2018 for a possible international partnership.

Precision Integrated Optics Near Infrared Imaging Experiment (PIONIER) Instrument capable of coupling four telescopes at the Paranal site (8.2 m or AT), forming and detecting fringes in the near infrared. Built as a prototype using optical fibres and integrated optics and implemented in 2010, this combiner is now used generally on the VLTI. Built by the Grenoble Observatory (France).

Probing the Universe with Advanced Optics (PUEO) Inspired by the Hawaiian word 'Pue'o' meaning 'owl'. First adaptive optics system based on the idea of analysing wave front curvature. Installed on the CFHT in Hawaii in 1996, it operated until 2012.

Rodrigue Near infrared camera with 64 pixels, equipping the ESO's 2.2 m telescope in La Silla in 1984 and built at the Paris Observatory.

SHARP2 Near infrared camera set up by the Max-Planck Institute für extraterrestriche Physik (Garching) on the ESO's 3.58 m telescope (NTT) in La Silla (Chile) in 1990. This camera was quite exceptional at the time, with 258×258 pixels. Replaced by the improved version SHARP2+, it was superseded 10 years later by the performance of the VLT.

Spectrograph for INtegral Field Observations in the Near Infrared (SIN-FONI) Integral field spectrograph operating in the near infrared and assisted by adaptive optics, SINFONI was set up at the Cassegrain focus of the telescope Yepun in 2005.

Spectro-Polarimetric High Contrast Exoplanet Research (SPHERE) Instrument set up at the Nasmyth focus of the telescope Yepun in Paranal in 2015. Designed for high resolution imaging with dedicated adaptive optics.

Sydney University Stellar Interferometer (SUSI) First generation optical interferometer set up in Narrabri (Australia) in 1983, operating in the visible on a 160 m baseline, with possible extension to 640 m.

Synthèse d'Ouverture en InfraRouge à Deux Télescopes (SOIRDETE) With a name inspired by the French for 'summer evening', this was an interferometer with two fixed 1.5 m telescopes on a 15 m baseline, using heterodyne fringe detection. It was set up on the Plateau de Calern near Grasse (France) in 1992 and operated until 1998.

Thirty Meter Telescope (TMT) Optical telescope of diameter 30 m, set up on Mauna Kea (Hawaii) by the California Institute of Technology as part of an international partnership. Scheduled for operation in the early 2020s.

TRIO Plan for an interferometer with 100 m baseline formed by three independent spaceborne telescopes, held in formation by a feedback system. Proposed by Antoine Labeyrie in 1984, this project was never realised, but triggered European reflection on a spaceborne optical interferometer.

Very Large Array (VLA) Radio interferometer comprising 27 mobile 25 m telescopes, with maximal possible baseline 36 km, built not far from Soccoro (New Mexico). Fully operational since 1980.

Very Large Telescope (VLT) Four 8.2 m telescopes and four 1.8 m telescopes, installed on Cerro Paranal in the Atacama Desert, Chile, since 1998. The 8.2 m telescopes can operate independently or in combinations as an interferometer or a larger light collector (instrument ESPRESSO). Set up by the European Southern Observatory (ESO).

Very Large Telescope Interferometer (VLTI) Particular mode of operation of the VLT in which at least two, but up to four of the eight telescopes are coupled to form an interferometer working in the visible, near infrared, or mid-infrared.

VIRGO Interferometric detector of gravitational waves, set up near Pisa (Italy) by a European partnership. It has been running since 2007, in partnership with LIGO in the United States.

VISible InfraRed Interferometer (VISIR) Optical interferometer designed by several European countries in the 1980s in preparation for the VLTI, but never actually pursued.

VLT Interferometer Commissioning Instrument (VINCI) First light instrument of the VLT interferometer, coupling two 8.2 m telescopes in 2001, then two auxiliary telescopes in 2005.

Yepun One of the four 8.2 m telescopes of the VLT. Also called UT4, its name refers to the star Sirius in Quechua.

Glossary

Accretion Accumulation of matter on a body. Accretion is caused by the gravitational attraction between the attracting body and the surrounding matter which accumulates on it.

Active galaxy Galaxy whose central region is an intense source of radiation at radio, infrared, visible, and X-ray wavelengths. This radiation is attributed to the accretion of matter by a massive black hole situated at the center of the galaxy, and called an active galactic nucleus (AGN).

Active optics Modification of an optical system by an external control mechanism. This refers in particular to control of the surface of a telescope's primary mirror to compensate for slow deformations from the ideal form giving the best images. These deformations may be due to the effects of wind, changes in orientation or temperature, and so on. Active optics differs from adaptive optics in that the latter acts a hundred to a thousand times faster to compensate for the different light travel times of light waves as they cross the Earth's atmosphere.

Acuity The capacity of the eye to make out details. The acuity of a normal eye is 20/20 and corresponds to a resolving power of about one minute of arc. If not corrected, myopia (short-sightedness) and hypermetropy (long-sightedness) reduce acuity. See also Resolution.

Adaptive optics Modification of an optical system by an externally imposed control system which compensates for degradation of the wave when it crosses the atmosphere. An adaptive system comprises a wave front analyser which measures the deformations of the wave front, a computer which calculates the adjustments that should be applied, and a deformable mirror which applies them. The whole system works by means of a closed feedback loop.

© Springer Nature Switzerland AG 2020
P. Léna, *Astronomy's Quest for Sharp Images*, Astronomers' Universe,
https://doi.org/10.1007/978-3-030-55811-6

Aether Medium supposed to allow the propagation of light, according to a hypothesis made at the beginning of the nineteenth century. The special theory of relativity eliminated the need for this assumption.

Airy disk Diffraction pattern observed in the image of a pointlike light source, for a telescope of given diameter. The central spot of light has an angular diameter in radians that is given approximately by the ratio of the light wavelength and the diameter of the telescope. It is surrounded by rings of light.

Angular resolution An optical system's capacity to produce an image of two pointlike light sources in which the images of the two points can be clearly distinguished from one another is specified by the minimum angular separation of the light sources for which this is possible. The resolution achieved in the image, i.e., the capacity to make out the details, is the resolution of the optical system which formed the image.

Aperture synthesis Method for reconstructing a complete image of an object, i.e., actually resembling the object, from distinct measurements of the fringes obtained by different pairs of telescopes. The contrasts and phases of the different sets of fringes are combined to calculate the final image.

Apparent diameter Angle subtended by an object at a distant observer.

Artificial star Creation of a light source in the Earth's stratosphere by illuminating the sodium atoms typically present at an altitude of about 80 km. This illumination is produced by a ground-based laser with a suitable wavelength.

Astrometry Method for precisely measuring the position of a body in the sky relative to some predefined coordinate system.

Black hole Spacetime singularity produced by such a great accumulation of matter that no known force internal to the matter will be able to resist its gravitational collapse. See also Chap. 7.

Brown dwarf Celestial object with mass somewhere between the minimal mass for an object to become a star $(0.08 M_{Sun})$ and the mass of a planet $(13 M_{Jupiter})$. A brown dwarf cannot trigger nuclear reactions in its core as a star would do, but they can occur briefly at its surface. Brown dwarfs are cold and emit very little light, but they can be observed in the near infrared.

Cassegrain focus The focal point where the image forms is given by the primary mirror after reflection on a secondary mirror. This focal point is located behind the primary mirror, which has a hole at its center to let the light through.

Charge coupled device (CCD) A photoelectric device detecting visible or near infrared light and comprising a great many pixels. Coupled with a computing electronic circuit, it can measure the light energy received at each point of an image. By digitisation, this circuit can then indicate the energy detected at each.

Coronagraph Device set up at the focal point of a telescope to mask a bright nearby object such as the Sun or a star, hence making it possible to view faint objects or regions in the neighbourhood.

Deformable mirror Thin, flexible mirror whose surface can be rapidly deformed by little motors called actuators placed behind the surface and controlled electrically.

This deformation advances or delays the reflected wave. The mirror may have tens, hundreds, or even thousands of actuators distributed regularly behind its surface.

Delay line Experimental device that can increase the light travel time of one light wave relative to another. Generally done using mirrors. The delay line can be fixed (constant delay) or mobile (controlled variable delay).

Diffraction Change of direction of a light wave when it encounters a material obstacle.

Doppler-Fizeau effect Change in the wavelength of wave emission received by an observer when the emitter is moving relative to the observer. The magnitude of the effect depends on the relative velocity and orientation of the motion.

Dual field Technique used in interferometry, whereby the interference fringes of two neighbouring objects are obtained simultaneously and can be compared.

Eddington limit When matter is accreted gravitationally by a massive body, it gains energy during its fall. This energy is transformed into heat and light radiation which then escapes toward the outside. The escaping radiation exerts a force called radiation pressure on the infalling matter, and this brakes its fall. A limit to the emitted light power is reached when the braking effect attains a value that prevents further accretion. This limiting emitted light power is the Eddington limit.

Exoplanet Planet orbiting another star than the Sun.

Field of view Region of space from which an optical device receives light and forms an image. This is two-dimensional and the position of a point in the field is specified by giving two angles.

Fringe tracking A set of interference fringes produced by a pair of telescopes is affected by propagation of the light from the celestial object through the atmosphere to each of the telescopes. The different light travel times along the two different paths vary very quickly and produce random shifts in the fringes. The amplitude of these shifts may correspond to several times the spacing within the fringes. The idea of fringe tracking is to measure the shift and compensate for it in real time using a delay line.

General relativity Mathematical and physical description of spacetime containing matter and energy.

Gravitation In Newtonian mechanics, an attractive force between any two masses, whose strength is proportional to the product of the two masses and inversely proportional to the square of the distance between them. The constant of proportionality G is the universal constant of gravitation, also called Newton's constant. In Einstein's general theory of relativity, gravity is understood rather as a modification of the geometry of spacetime by the presence of matter or energy. The curvature of spacetime induced in this way determines the trajectory of light or matter.

Gravitational wave Perturbation of spacetime caused by a change in the position of matter within it. This perturbation propagates at the speed of light in the form of a wave, carrying energy and able to cause other matter to move when it encounters it along its path.

Habitable zone Region between a minimal and a maximal distance from a star of given luminosity such that the radiative equilibrium temperature at the surface of a planet located within this region would lie between 0 and 100°C. The idea is then that there could be liquid water at the surface of the planet.

Image sharpness See Resolution.

Infrared region Region of the light spectrum covering wavelengths 0.7–5 μm (near infrared), 5–20 μm (mid-infrared), and 20–200 μm (far infrared). The definition of these boundaries is relatively arbitrary.

Integrated optics Optical system using miniaturised components and microelectronics.

Intensity interferometry Particular interference mode in which it is not the interference fringes that are observed, but rather the correlation between the arrivals of individual photons on distinct telescopes.

Interference Phenomenon occurring when two light waves with definite relative phase are superposed. The resulting wave exhibits periodic spatial variations of the light intensity, with maxima and minima.

Interference fringes Patterns of alternating light and darkness. Fringes can be straight, circular, or other. They are formed by the superposition of light waves (interference) in the special situation where these waves are totally or partially coherent. See also Interference.

Interferometer Experimental device exploiting interference between two or more light waves.

Nasmyth focus Focal point located to the side of the telescope tube in the case of an altazimuth mount. This facilitates the installation of heavy or bulky instrumentation.

Occultation When one celestial object passes in front of another, entirely covering the latter. The classic case is the occultation of a star by the Moon or an asteroid.

Optical fibre Very fine cylindrical tube of transparent material in which light can propagate great distances without excessive attenuation. Single-mode fibres conserve the phase of the propagating light wave.

Optical interferometry Interferometer operating in the optical region (visible or infrared wavelengths).

Optical region Rather loosely defined region of the light spectrum covering visible and infrared wavelengths.

Optical trombone Device for modifying the light travel time by changing the path length. See also Delay line.

Phase closure Technique used in interferometry to cancel out phase variations produced in the Earth's atmosphere along the different paths the light takes to reach the telescopes of the interferometer.

Phase difference The different propagation times of light along two different paths leads to a relative shift in the oscillations of the light signals, i.e., a difference in their respective phases.

Primary mirror The main mirror of a telescope, with a concave shape, receiving the light from the celestial object under investigation and its surroundings. The primary mirror sends the light to other mirrors which guide it to the focal point or points where images are formed.

Quasar Extremely remote celestial body that is the source of intense emissions. Due to the distance, it appears almost pointlike to the telescope, whence the name 'quasi-star'. Quasars are located at the center of active galaxies and their radiation is due to the accretion of matter by a black hole.

Radian Unit of angle (rad), defined with reference to a circle of unit radius as the angle subtended at the center by a length of one unit on the circumference. Hence, $360° = 2\pi$ radians and 1 radian is therefore approximately equal to $57.29°$. One second of arc (1 arcsec) is equal to $1/(57.29 \times 60 \times 60) \approx 5 \times 10^{-6}$ rad.

Radio region Region of the light spectrum covering all wavelengths beyond about $500\,\mu$m, generally divided into submillimeter, millimeter, centimeter, and so on.

Reflecting telescope Any telescope in which the image is formed after reflection on the primary mirror.

Refracting telescope Any telescope in which the image is formed by a converging (convex) refracting lens which receives the light from the celestial object, as in most photographic devices.

Resolving power Angle measuring the capacity of an optical system, such as the eye or a telescope, to distinguish two pointlike light sources. See also Resolution or Acuity.

Schwarzschild radius Length characterising a black hole of a given mass. This radius is proportional to the mass of the black hole. Indeed, it depends only on the mass and two universal constants of physics, viz., the speed of light c and the universal gravitational constant G.

Scintillation Rapid changes in brightness, i.e., light intensity, of a star, as observed by the naked eye when the light crosses non-uniform layers of air, with different temperatures and hence different refractive indexes. Commonly known as twinkling.

Seeing Effect due to the propagation of light through layers of different temperature in the atmosphere. The image of a star given by a telescope will thus be a blurred disk, i.e., spread out and fluctuating in time, known as the seeing disk.

Seismology Study of acoustic waves propagating in a solid body like the Earth or gaseous body like the Sun or another star.

Sensitivity Energy level of a signal, e.g., a light signal, below which detection is no longer possible. The signal may then be swamped by interference effects inherent in the detection method, known as noise. A measurement instrument can be characterised by its sensitivity to weak signals.

Spatial frequency Quantity characterising a structure which repeats itself identically in space at regularly spaced positions.

Speckle Fine detail appearing within the seeing disk when it is observed over a very short time, thereby freezing the effects of the Earth's atmosphere on the light as it travelled through it. The level of detail is determined by the size of the telescope's Airy disk.

Transit Passage of an exoplanet in front of or behind the disk of its host star, as seen from the Earth. Passage in front of the star is known as a primary transit, and behind the star as a secondary transit.

Transition disk Disk of matter (gas and dust) orbiting around another star and temporarily present during the formation of the star and its planets.

Turbulence State of motion of a fluid characterised by a random distribution of the fluid velocity at each point within it and at each instant of time.

Velocimetry Indirect method for detecting the presence of an exoplanet in orbit around a host star. The velocity of the star is determined relative to the Earth-based observer by measuring the shift in the star's spectral lines due to the Doppler effect. Any periodic variations in this velocity are interpreted as being due to the effect of the force of gravity caused by a planet as it orbits around the star, since this will affect the motion of the star.

Very long baseline interferometry (VLBI) An interferometry configuration used in radioastronomy in which telescopes several thousand kilometers apart can be coupled to form interference fringes. The resolution of the image produced by aperture synthesis is given in radians by the ratio of the wavelength and the length of the maximal baseline.

Wave front Set of points, constituting a surface, where the oscillations of a light wave are exactly in phase with each other. When a stone is dropped into a pool of water, causing waves to propagate over the surface, the set of points at the same height on a given circular ripple corresponds to a wave front. For sound waves or light waves propagating from a point source in three-dimensional space, the wave fronts will be spherical surfaces. Far from the source, the curvature of the wave front will become negligible locally, and the wave front detected from a star, for example, will be almost planar.

Wave front analyser Device for very rapidly measuring, at a certain number of points, the deviations of a wave front from a plane when it has been distorted. These deviations are generally expressed as a fraction or multiple of the wavelength.

Bibliography

J.P. Angel, Discussion Notes, p. 59, in *Scientific Importance of High Angular Resolution at Infrared and Optical Wavelengths*, ed. by M.H. Ulrich, K. Kjär, Garching (1981)

J. Arnould, *Turbulences dans l'Univers. Dieu, les extraterrestres et nous* (Albin Michel, Paris, 2017)

H.H. Aumann, IRAS observations of matter around nearby stars. Publ. Astron. Soc. Pac. **97**, 885 (1985)

H. Babcock, The possibility of compensating atmospheric seeing. Publ. Am. Soc. Pac. **65**, 229 (1953)

J. Beckers, Adaptive Optics for astronomy: principles, performance and applications. Ann. Rev. Astr. Ap. **31**, 13–92 (1993)

J.M. Beckers, Interferometric imaging in astronomy: a personal retrospective, in *The Sun and Planetary Systems. Paradigms for the Universe*. Rev. Mod. Astron. **17**, 239 (2004)

S. Beckwith, B. Zuckerman, M.F. Skrutskie, H.M. Dyck, Discovery of halos of solar system size around young stars. Ap. J. **287**, 793 (1984)

C. Bertout and J. Bouvier, Interferometric imaging of protoplanetary discs around young stars, in *High Resolution Imaging by Interferometry*, vol. 1, ESO Proceedings (1988), p. 69

J.-L. Beuzit, N. Hubin, ADONIS, a user friendly system. ESO Messenger **71**, 528 (1993)

J.-L. Beuzit et al., SPHERE: the exoplanet imager for the Very Large Telescope (2019). https://arxiv.org/abs/1902.04080

P. Binetruy, *Gravity. The Quest for Gravitational Waves* (Oxford University Press, Oxford, 2018)

A. Blaauw, *ESO's Early History. The European Southern Observatory from Concept to Reality* (ESO Garching, München, 1991)

© Springer Nature Switzerland AG 2020
P. Léna, *Astronomy's Quest for Sharp Images*, Astronomers' Universe,
https://doi.org/10.1007/978-3-030-55811-6

D.C. Black, M.S. Matthews (eds.), *Protostars & Planets II*. Space Science Series (University of Arizona Press, Tucson, 1985)

L. Blanchet, *Les ondes gravitationnelles*. Reflets de la Physique, vol. 52 (2017)

A. Boccaletti, A. Lagrange et al., Independent confirmation of imaging of Beta Pictoris b with NICI. Astron. Ap. **551**, L14 (2013)

D. Bonneau, *Mieux voir les étoiles. 1er siècle de l'interférométrie stellaire* (EDP Sciences, Les Ulis, 2019)

R.-M. Bonnet, L. Woltjer, *Surviving 1000 Centuries. Can We Do It?* (Springer, Berlin, 2008)

G. Bourdet, in *ESO Conference on Optical Telescopes of the Future*. ESO Proceedings (Garching, 1978)

W. Brandner et al., NAOS + Conica at Yepun: first VLT adaptive optics system sees first light. ESO Messenger **107**, 1–6 (2002)

A. Broderick, A. Loeb, Portrait of a black hole, Scientific American, Special issue 22 (2013)

S. Brunier, A.-M. Lagrange, *Great Observatories of the World* (Firefly Books, Richmond Hill, 2005)

D. Charbonneau et al., Detection of planetary transits across a Sun-mass star. Ap. J. **529**, L45–L48 (2000). https://fr.wikipedia.org/wiki/HD_209458_b

G. Chardin, *L'insoutenable gravité de l'Univers* (Le Pommier, New York, 2018)

G. Charpak, P. Léna, Y. Quéré, *L'Enfant et la science* (O. Jacob, Paris, 2005)

G. Chauvin, A. Lagrange et al., A giant planet candidate near a young brown dwarf. Direct VLT/NaCo observations using IR wavefront sensing. Astron. Astrophys. **425**, L29 (2004)

G. Chauvin, A. Lagrange et al., Giant planet companion to 2MASSW J1207334-393254. Astron. Astrophys. **438**, 25–28 (2005)

A. Chelli, P. Léna, F. Sibille, Angular dimensions of accreting young stars. Nature **278**, 143 (1979)

M.K. Crawford, R. Genzel, C. Townes, et al., Mass distribution in the galactic center. Nature **315**, 467–470 (1985)

F. Crifo et al., Beta Pictoris revisited by Hipparcos. Star properties. Astron. Astrophys. **320**, L29–L32 (1997)

J. Davis, J.W. Tango, New determination of the angular diameter of Sirius. Nature **323**, 234 (1986)

J.M. de Heredia, in *Les Conquérants* (Les trophées, Paris, Gallimard, 1981) [1893]

A. de Saint Exupéry, *Wind, Sand and Stars* (Harcourt, New York, 1939)

H.J. Deeg, J.A. Belmonte, (eds.), *Handbook of Exoplanets* (Springer, Berlin, 2018)

D. Defrère, A. Léger, O. Absil, Space-based interferometry to study exoplanetary atmospheres. Exp. Astr. **46**, 543 (2018). https://arxiv-org.ezproxy.obspm.fr/abs/1801.04150

H. Dole, *Le côté obscur de l'univers* (Dunod, Paris, 2017)

H.M. Dyck, Near-infrared slit-scans of molecular cloud sources. Astr. J. **85**, 891 (1980)

H.M. Dyck, E. Kibblewhite, Giant infrared telescopes for astronomy—a scientific rationale. Publ. Astron. Soc. Pac. **98**, 260 (1986)

A. Eckart, R. Genzel, Observations of stellar proper motions near the Galactic center. Nature **383**, 415–417 (1996)

EHT Collaboration, First M87 Event Horizon Telescope Results. I. The shadow of the supermassive black hole in the center of the M87 galaxy. Astrophys. J. **875**, L1–17 (2019)

A. Einstein, Annalen der Physik **35**, 898–908 (1911)

A. Einstein, *Œuvres choisies* vol. 6, presented by F. Balibar, O. Darrigol, and B. Jech, Le Seuil & édns du CNRS (1989)

F. Eisenhauer, GRAVITY. The AO-assisted, two-object beam-combiner, in *The Power of Optical/IR Interferometry* (Springer, Berlin, 2006)

F. Eisenhauer et al., GRAVITY: observing the universe in motion, ESO Messenger **143**, 16–24 (2011)

D. Enard, The VLT: genesis of a project. ESO Messenger **50**, 30–32 (1987)

T. Encrenaz, *La montagne magique. Mauna Kea. Hawaii* (S. Fischer Verlag, Berlin, 2018). ISBN 978-2-955385-0-2

S. Ertel, O. Absil et al., A near-infrared interferometric survey of debris-disk stars. IV. An unbiased sample of 92 southern stars observed in H band with VLTI/PIONIER. Astron. Astrophys. **570**, A128 (2014). https://doi.org/10.1051/0004-6361/201424438, arXiv:1409.6143

C. Fehrenbach, *Des hommes, des télescopes et des étoiles* (Vuibert, Paris, 1990)

H. Fizeau, P. Bordin. Rapport sur le concours pour l'année 1867. C.R.A.S. **66**, 932 (1868)

M. Gargaud et al. (eds.), *From Suns to Life. A Chronological Approach to the History of Life on Earth* (Springer, Berlin, 2006)

J. Gay, A. Journet, Infrared interferometry. Nature **241**, 32–22 (1973)

E. Gendron, Doctoral Thesis: Optimisation de la commande modale en optique adaptative: applications à l'astronomie (1995). https://hal.archives-ouvertes.fr/tel-01418424

R. Genzel, R. Schödel, T. Ott, A. Eckart, T. Alexander, F. Lacombe, D. Rouan, B. Aschenbach, Near-infrared flares from accreting gas around the supermassive black hole at the Galactic Centre. Nature **425**, 934 (2003)

D. Gezari, A. Labeyrie, R. Stachnik, Speckle interferometry: diffraction-limited measurements of nine stars with the 200-inch telescope. Ap. J. **173**, L1 (1972)

A. Ghez et al., High proper-motion stars in the vicinity of Sagittarius A*: evidence for a supermassive black hole at the center of our galaxy. Ap. J. **309**, 678 (1998)

R. Gratton et al., Searching for the near-infrared counterpart of Proxima c using multi-epoch high-contrast SPHERE data at VLT. Astron. Astrophys. **638**, (2020). arXiv200406685G/abstract

Gravity Collaboration, First light for GRAVITY: phase referencing optical interferometry for the Very Large Telescope Interferometer. Astron. Astrophys. **602**, A94 (2017)

Gravity Collaboration, Detection of the gravitational redshift in the orbit of the star S2 near the Galactic center massive black hole, Astron. Asrophys. **615**, L15 (2018)

Gravity Collaboration, Detection of orbital motions near the last stable circular orbit of the massive black hole SgrA*. Astron. Astrophys. **618**, L10 (2018)

F.M. Grimaldi, *Physico-mathesis de lumine, coloribus, et iride, aliisque adnexis libri duo: opus posthumum.* Ex Typographia haeredis Victorii Benatii, impensis Hieronymi Berniae, publ. (1665)

A. Hatzes et al., A planetary companion to Gamma Cephei A. Ap. J. **599**, 1383 (2003)

S.R. Heap, Bull. Am. Astr. Soc. **191**, 4702 (1997)

S.-S. Huang, The problem of life in the Universe and the mode of star formation. Publ. Astron. Soc. Pac. **7**(422), 421–424 (1959)

S. Issaoun et al., The size, shape, and scattering of SgrA* at 86 GHz: First VLBI with ALMA. Astrophys. J. **871**, 30 (2019)

M. Keppler, M. Benisty et al., Discovery of a planetary mass companion within the gap of the transition disc around PDS 70. Astron. Astrophys. **617**, A44 (2018). https://arxiv.org/abs/1806.11568

P. Kervella, *Application de l'interférométrie à l'étude des Céphéides et des étoiles naines*, Research Qualification, University Paris-Diderot and Paris Observatory (2007)

J. Kobus et al., The potential of combining MATISSE and ALMA observations: constraining the structure of the innermost region in protoplanetary discs. Astron. Astrophys. **622**, A47 (2019)

B. Koehler, S. Lévêque, P. Gitton, A decade of VLTI technical development, in *Interferometry for Optimal Astronomy II*, ed. by W. Traub, vol. 4838 (SPIE, Bellingham, 2003), pp. 846–857

S. Kraus, G. Weigelt et al., Tracing the dynamic orbit of the young, massive, high eccentricity binary system θ_1 Orionis C. First results from VLTI aperture-synthesis imaging and ESO 3.6 m visual speckle interferometry. ESO Messenger **136**, 44–47 (2009)

E.S. Kulagin, Measurements of Capella with the Pulkovo stellar interferometer. Sov. Astron. **14**, 445 (1970)

A. Labeyrie, Attainment of diffraction limited resolution in large telescopes by Fourier analysing speckle patterns in star images. Astr. Ap. **6**, 85 (1970)

A. Labeyrie, Interference fringes obtained on VEGA with two optical telescopes. Ap. J. **196**, L71 (1975)

S. Lacour et al., First direct detection of an exoplanet by interferometry. K band spectroscopy of HR8799 e. Astron. Astrophys. **623**, L11 (2019)

A.-M. Lagrange, *Exoplanètes. Méthodes de détection.* Encyclopedia Universalis (2019). https://www.universalis.fr/encyclopedie/exoplanetes/

A.M. Lagrange, G. Chauvin, Beta Pictoris, a laboratory for planetary formation studies. ESO Messenger **150**, 39–43 (2012)

A.-M. Lagrange, P. Léna, *Exoplanètes ou planètes extra-solaires.* Encyclopædia Universalis. Accessed 23 January 2019. www.universalis.fr/encyclopedie/exoplanetes-ouplanetes-extrasolaires/

A.M. Lagrange et al., A giant planet imaged in the disk of the young star Beta Pictoris. Science **329**, 57 (2010)

O. Lai et al., FlyEyes: a CCD-based wavefront sensor for PUEO, the CFHT curvature AO system. Publ. Astron. Soc. Pac. **123**, 902 (2011)

P.R. Lawson (ed.), *Selected Papers on Long Baseline Stellar Interferometry*. SPIE Milestone Series, vol. MS 139 (SPIE The International Society for Optical Engineers, Paris, 1997)

J.-B. Le Bouquin, J.-P. Berger et al., PIONIER, a 4-telescope visitor instrument at VLT. Astr. Ap. **535**, A67 (2011)

A. Léger, J.M. Mariotti, B. Mennesson, M. Ollivier, J.L. Puget, D. Rouan, J. Schneider, The DARWIN project, an infrared nulling interferometer in space. Astrophys. Space Sci. **241**, 135–146 (1996)

A. Léger, D. Rouan, J. Schneider, Transiting exoplanets from the Corot Space mission. Astr. Ap. **506**, 287–302 (2009)

P. Léna, Jean-Marie Mariotti. ESO Messenger **93**, 47 (1998)

P. Léna, Adaptive Optics: Astronomical results and perspectives. Exp. Astron. **7**, 281 (1997)

P. Léna, *La photographie astronomique*, in *Éclats d'histoire*. Actes Sud/Institut de France (2003)

P. Léna, *Racing the Moon's Shadow with Concorde 001* (Springer, Berlin, 2015)

J. Lequeux, *Hippolyte Fizeau, physicien de la lumière* (EDP Sciences, Les Ulis, 2014)

J. Lequeux, *François Arago. A 19th Century French Humanist and Pioneer in Astrophysics* (Springer, Berlin, 2016)

J. Lequeux, T. Encrenaz, *La révolution des exoplanètes* (EDP Sciences, Les Ulis, 2017)

J. Lequeux, T. Encrenaz, *Des planètes terrestres aux exoplanètes habitables* (EDP Sciences, Les Ulis, 2019)

M. Lesieur, *Turbulence in Fluids* (Springer, New York, 2008)

K.Y. Lo et al., On the size of the galactic center compact radio source: diameter < 20 AU. Nature **315**, 124 (1985)

S. Loiseau, G. Perrin, *Interférométrie optique: ombres et lumières dans l'univers* La Recherche, vol. 292 (1996)

C. Lovis et al., Atmospheric characterization of Proxima b by coupling the SPHERE high-contrast imager to the ESPRESSO spectrograph. Astron. Astrophys. **599**, A16 (2017). arXiv:1609.03082

J.-P. Luminet, *Le Destin de l'univers. Trous noirs et énergie sombre* (Fayard, Paris, 2010)

D. Lynch, W. Livingston, *Colour and Light in Nature*, 2nd edn. (Cambridge University Press, Cambridge, 2001)

C. Madsen, *The Jewel on the Mountaintop* (Wiley, Hoboken, 2012)

B. Maitte, *La Lumière* (Poche Seuil, Paris, 2005)

J.-M. Mariotti, V. Coudé du Foresto, G.S. Perrin, P. Léna, *Interferometric connection of large telescopes at Mauna Kea, Proceedings of SPIE*. Astronomical Interferometry, vol. 3350 (1998)

C. Marois, B. Zuckerman, Q.M. Konopacky, B. Macintosh, T. Barman, Images of a fourth planet orbiting HR8799. Nature **468**, 7327, 1080–1083 (2010)

M. Mayor, P.-Y. Frei, *Les nouveaux mondes du cosmos: à la découverte des exoplanètes* (Seuil, 2001)

M. Mayor, P.-Y. Frei, *New Worlds in the Cosmos. The Discovery of Exoplanets* (Cambridge University Press, Cambridge, 2003)

A. Mérand, The VLTI roadmap. ESO Messenger **171**, 14 (2018)

A. Mérand et al., GRAVITY science verification. ESO Messenger **170**, 16–19 (2017)

F. Merkle, (ed.), *NOAO–ESO Conference on High-Resolution Imaging by Interferometry.* ESO Proceedings, vol. 29 (1988)

F. Merkle, P. Kern, P. Léna, F. Rigaut, J.-C. Fontanella, G. Rousset, C. Boyer, J.-P. Gaffard, P. Jagourel, Successful tests of adaptive optics. Messenger **58**, 1–4 (1989)

A. Michelson, Measurement of Jupiter's satellites by interference. Publ. Astr. Soc. Pac. **3**, 274 (1891)

J.-D. Mollon, The origin of the concept of interference. Philos. Trans. Lond. Roy. Soc. A **360**, 807–819 (2002)

J. Monnier et al., Planet Formation Imager (PFI): introduction and technical considerations (2014). arXiv:1407.7032v1

J. Monnier, S. Kraus, M. Ireland et al., The planet formation imager. Exp. Astron. **46**, 517 (2018)

T. Montmerle, D. Ehrenreich, A.-M. Lagrange (eds.), *Physics and Astrophysics of Planetary Systems* (EDP Sciences, Les Ulis, 2010)

D. Mouillet, A.M. Lagrange, J.-L. Beuzit, A stellar coronograph for the ComeOn+ adaptive optics system. Astron. Astrophys. **324**, 1083 (1997a)

D. Mouillet et al., A planet on an inclined orbit as an explanation of the warp in the Beta Pictoris disc. Month. Not. R. Astron. Soc. **292**, 896 (1997b)

D. Mourard et al., GI2T and Delta Ceph. Astron. Astrophys. **317**, 789–792 (1997)

D. Mozurkewich, M. Shao, M. Colavita, et al., Phase referenced averaging as a method for decreasing the variance of visibility measurements, in *High Resolution Imaging by Interferometry*, ed. by F. Merkle. ESO Proceedings, vol. II (1988)

A. Müller, M. Keppler et al., Orbital and atmospheric characterization of the planet within the gap of the PDS 70 transition disc. Astron. Astrophys. **617**, L2 (2018). https://arxiv.org/abs/1806.11567

R. Narayan, Sparks of interest. Nature **425**, 908 (2003)

M. Nowak et al., Beta Pic c, a hot-start planet at 2.7 astronomical units. Astron. Astrophys. (in press, 2020)

B. Pascal, Pensées, Disproportion de l'homme, in *Œuvres complètes*, vol. 72 (La Pléiade Gallimard, Paris, 1954)

B. Pascal, *Œuvres complètes* (La Pléiade, Genève, 1957), p. 359ff

G. Perrin, V. Coudé du Foresto, S.T. Ridgway, J.-M. Mariotti, W.A. Traub, N.P. Carleton, Extension of the effective temperature scale of giants to types later than M6. Astron. Astrophys. **331**, 619–626 (1998)

G. Perrin, S.T. Ridgway, B. Mennesson, et al., Unveiling Mira stars behind the molecules. Confirmation of the molecular layer model with narrow band near-infrared interferometry. Astron. Astrophys. **426**, 279 (2004)

G. Perrin, J. Woillez, O. Lai, et al., Interferometric coupling of the Keck telescopes with single-mode fibers. Science **311**, 194 (2006)

D. Proust, J. Schneider, *Où sont les autres? À la recherche d'une vie dans l'univers* (Seuil, Paris, 2007)

L.F. Richardson, *Weather Prediction by Numerical Processes* (Cambridge University Press, Boston, 1922), p. 66

A. Richichi, F. Delplancke, F. Paresce, A. Chelli (eds.), *The Power of Optical/IR Interferometry: Recent Scientific Results and 2nd Generation Instrumentation* (Springer, Berlin, 2008). Summarised in ESO Messenger **120**, 48–51 (2005)

F. Rigaut, G. Rousset, P. Kern, J.-C. Fontanella, J.-P. Gaffard, F. Merkle, P. Léna, Adaptive optics on a 3.6 meter telescope: results and performances. Astron. Astrophys. **250**, 280 (1991)

F. Roddier, The effect of atmospheric turbulence in astronomy. Prog. Opt. **19**, 281 (1981)

F. Roddier (ed.), *Adaptive Optics in Astronomy*, 2nd edn. (Cambridge University Press, Cambridge, 2004)

F. Roddier et al., Adaptive optics imaging of GG Tau: optical detection of the circumbinary ring. Astophys. J. **463**, 326 (1996)

A.E. Rogers et al., Small-scale structure and position of Sagittarius A* from VLBI at 3 millimeter wavelength. Astrophys. J. **434**, L59 (1994)

D. Rouan, Caravelle 116. Osiris et Astronomie: l'astronomie infrarouge aéroportée en France, L'Astronomie no. 62, Société astronomique de France (2013)

D. Rouan, A. Baglin, The Exosolar Planets Program of the Corot satellite. Earth Moon Planets **81**(1), 79–82 (1998)

D. Rouan et al., Near-IR images of the torus and micro-spiral structure in NGC 1068 using adaptive optics. Astron. Astrophys. **339**, 687 (1998)

G. Rousset, J.-C. Fontanella, P. Kern, P. Léna, P. Gigan, F. Rigaut, J.-P. Gaffard, C. Boyer, P. Jagourel, F. Merkle, First diffraction limited astronomical images obtained with adaptive optics. Astron. Astrophys. **230**, L29 (1989)

M. Schlenker, Michel Soutif (1921–2016). Reflets de la Physique **51**, 44 (2016)

J. Schneider, Future exoplanet research: science questions and how to address them, in *Handbook of Exoplanets*, ed. by H.J. Deeg, J.A. Belmonte (Springer, Berlin, 2018), pp. 3245–3267

R. Schödel et al., A star in a 15.2-year orbit around the supermassive black hole at the center of the Milky Way. Nature **419**, 694–696 (2002)

R. Schödel et al., Stellar dynamics in the central arcsec of our Galaxy. Astrophys. J. **596**, 1015 (2003)

K. Schwarzschild, Über das Gravitationelfeld eines Massenpunktes nach der Einsteinchen Theorie, Sitzungsberichte der Königlich Preußischen Akademie der Wissenschaften (Berlin) (1916), pp. 189–196

S. Seager, Exoplanet habitability. Science **340**, 577–581 (2013)

G. Setti, G.G. Fazio (eds.), *Infrared Astronomy* (Springer, Berlin, 1978)

M. Shao, M. Colavita, Long baseline optical and infrared stellar interferometry. Ann. Rev. Astron. Astrophys. **30**, 457 (1992)

M. Shao, M. Colavita, Potential of long-baseline infrared interferometry for narrow-angle astrometry. Astron. Astrophys. **262**, 353–358 (1992)

D. Simons, J.-P. Maillard, Fourier imaging spectroscopy of the Galactic Center. Publ. Astr. Soc. Pac. **102**, 232 (1996)

E. Stephan, Sur les franges d'interférence observées avec de grands instruments dirigés sur Sirius et sur plusieurs autres étoiles; conséquences qui peuvent en résulter, relativement au diamètre angulaire de ces astres. C. R. Acad. Sci. **76**, 1008 (1873)

D. Stone, *Einstein and the Quantum. The Quest of the Valiant Swabian* (Princeton University Press, Princeton, 2013)

J.F. Stuyck-Taillandier, CNRS's European Policy from 1988 to 1994. Histoire de la recherche contemporaine **1**(1), 26–35 (2012). https://journals.openedition.org/hrc/197

C. Thom, P. Granés, F. Vakili, Optical interferometry measurements of γ Cassiopeiae's envelope in the Hα line. Astron. Astrophys. **165**, L13 (1986)

D.E. Thomsen, Taking the measure of the stars. Sci. News **131**, 10 (1987)

W. Tobin, J. Lequeux, *Léon Foucault* (EDP Sciences, Les Ulis, 2002)

M.H. Ulrich, K. Kjär, (eds.), *Scientific Importance of High Angular Resolution at Infrared and Optical Wavelengths*, 24–27 March 1981 (European Southern Observatory, Garching, 1981)

F. Vakili, S. Loiseau (eds.), *Interférométrie visible et IR dans l'espace.* Forum at the Paris Observatory, PNHRA INSU-CNRS (1995)

F. Vakili et al., Evidence for one-armed oscillations in the equatorial disk of zeta Tauri from GI2T spectrally resolved interferometry. Astron. Astrophys. **335**, 261–264 (1998)

G. van Belle, From PTI to the Keck Interferometer, in *VLT Opening Symposium* (Springer, Berlin, 1999)

G.T. van Belle, The VLTI Prima facility. ESO Messenger **134**, 6 (2008)

F.H. Vincent, T. Paumard, G. Perrin, L. Mugnier, F. Eisenhauer, S. Gillessen, Performance of astrometric detection of a hotspot orbiting on the innermost stable circular orbit of the galactic center black hole. Mon. Not. Roy. Soc. **412**, 2653 (2011)

O. von der Lühe, A new plan for the VLTI. ESO Messenger **87**, 8–14 (1997)

R.S. Westfall, *The Life of Isaac Newton* (Cambridge University Press, Cambridge, 1994)

R.N. Wilson, The history and development of the ESO active optics system. ESO Messenger **113**, 5 (2003)

A. Wolszczan, Two planets around a 6.2-ms pulsar PSR 1257+12? Bull. AAS **23**, 1347 (1991)

L. Woltjer, *Europe's Quest for the Universe* (EDP Sciences, Paris, 2006)

N. Woolf, R. Angel, in *Optical and Infrared Telescopes for the 1980s*, ed. by A. Hewitt (Kitt Peak National Observatory, Tucson, 1980), p. 1062

G. Zins et al., PIONIER: a four-telescope instrument for the VLTI. ESO Messenger **146**, 12–17 (2011)

Index

A

Absil, O., 182
Airy, G.B., 26, 81
Al Aitham, 19
Allende, S., 120
Angel, R., 96, 136, 228
Aquinas, T., 159
Arago, F., 14, 16, 63
Archimedes, 42
Ardeberg, A., 2, 121
Aristotle, 15
Arnould, J., 159

B

Babcock, H., 34, 36, 41, 42, 45, 49, 73, 218, 223
Bachelet, M., 227
Bacon, R., 19
Baglin, A., 170
Baldwin, J., 96, 148–150
Baranne, A., 11
Barish, B.C., 192
Barnard, E.E., 160
Beckers, J., 113, 123–128, 133, 147, 206
Beckwith, S., 138

Benisty, M., 154
Berger, J.-P., 154, 224, 231
Besso, M., 184
Beuzit, J.-L., 133, 173
Blazit, A., 74
Bonneau, D., 74, 75, 136
Bonner, E., 52
Bosc, I., 93
Bourdet, G., 45, 46, 223
Bourlon, P., 122
Boyer, C., 51
Brecht, B., 20
Brillet, A., 192
Bruno, G., 67
Burke, B., 163

C

Césarsky, C., 128, 131, 138
Chaffee, F., 155
Chalabaev, A., 52
Charbonneau, D., 171, 178
Chauvin, G., 166
Chelli, A., 38, 40
Chirac, J., 107, 121
Chou En-Lai, 114

© Springer Nature Switzerland AG 2020
P. Léna, *Astronomy's Quest for Sharp Images*, Astronomers' Universe,
https://doi.org/10.1007/978-3-030-55811-6